Industrial Applications
of Holography

Industrial Applications of Holography

Edited by
JEAN ROBILLARD
New Mexico State University

H. JOHN CAULFIELD
University of Alabama

New York Oxford
OXFORD UNIVERSITY PRESS
1990

Oxford University Press

Oxford   New York   Toronto
Delhi   Bombay   Calcutta   Madras   Karachi
Petaling Jaya   Singapore   Hong Kong   Tokyo
Nairobi   Dar es Salaam   Cape Town
Melbourne   Auckland

and associated companies in
Berlin   Ibadan

Copyright © 1990 by Oxford University Press, Inc.

Published by Oxford University Press, Inc.,
200 Madison Avenue, New York, New York 10016

Oxford is a registered trademark of Oxford University Press

All rights reserved. No part of this publication may be reproduced,
stored in a retrieval system, or transmitted, in any form or by any means,
electronic, mechanical, photocopying, recording, or otherwise
without the prior permission of Oxford University Press.

Library of Congress Cataloging-in-Publication Data
Industrial applications of holography/edited by Jean Robillard
and H. John Caulfield
p.   cm.
Papers from a symposium on "Industrial Uses of Holography" held Oct. 1987
at the Applied Optics Laboratories of New Mexico State University.
Includes bibliographies and indexes.
ISBN 0-19-505855-0
1. Holography—Industrial applications—Congresses.
I. Robillard, Jean.   II. Caulfield, H. J. (Henry John), 1936–
TA1542.I53   1990
621.36'75—dc20   89-9453   CIP

9 8 7 6 5 4 3 2 1

Printed in the United States of America
on acid-free paper

Foreword

Since holography was reinvented and pushed forward by the introduction of the laser in the early 1960s, the technique has grown from an optical curiosity into an important technical tool. Millions of people are confronted (many unknowingly) with holographic images as they are becoming common sights on books and credit cards. It is, however, interesting to note that even for holographic old-timers, the fascination with the magical image never seems to end.

As a technique, holography spans several professional fields and the ideal holographer should really be proficient within optics, mechanics, and chemistry (new application areas also demand use of electronics, video systems, computers, and image processing). Still conferences on holography have a tendency to concentrate on one aspect of the technique. This specialization may be welcome to researchers within that particular area, but it does not provide the overview often desired by the end user. Worse, however, it does not give the cross-information between specialists working in different branches that is so necessary for new ideas and further development within science.

The "International Symposium on Industrial Uses of Holography" in Las Cruces, New Mexico succeeded very well in meeting the two last demands. It brought together top specialists within key areas in the techniques and applications of holography, which provided up-to-date critical reviews of their fields and pointed out possible future developments. The atmosphere of presentation was relaxed with a liberal time schedule that encouraged an unusually high and fruitful level of interaction between the speakers and the audience. In addition to the invited papers, users of the holographic techniques presented their results and problems, which again stimulated lively discussions. The organization and running of the conference went smoothly thanks to the hard work of the local committee.

In my opinion the conference was very successful and definitively deserves a prime niche in the maze of holographic conferences. It is my hope that it can be repeated at regular intervals so that a larger audience can reap the fruits of its content.

Last I would like, on behalf of the the invited foreign speakers, to express my deepest thanks to the local committee for offering us an unusual hospitality in the true spirit of the friendly Southwest.

OLE J. LØKBERG
Chairman

Preface

In October 1987, a symposium on "Industrial Uses of Holography" was held at the Applied Optics Laboratories of New Mexico State University. The purpose of this symposium was to stimulate interactions between scientists concerned with industrial applications of holography.

The papers presented at the symposium covered a wide spectrum of applications including nondestructive testing, recording systems, computer generated holograms, data processing, holographic recording materials, as well as the teaching of holography in engineering.

As could be expected, nondestructive testing provided the most important part of the papers. From very special subjects, such as high temperature strain measurements and underwater inspection to the rapidly expanding use of TV holography, the authors also contributed to submicron inspection for microelectronics and strain detection in solar cells.

A most stimulating paper on "holographic virtual interconnections" with possible application to data processing provided a futuristic note to the Symposium. A sophisticated yet simple recording head using holographic optical elements (HOP) for optical discs was an example of Japanese high technology applied to a very current problem.

The display of computer-generated holographic images using a liquid crystal spatial light modulator was another interesting topic which showed how holograms can be virtually processed without recording substrate.

Real-time recording materials for holography, a subject of current interest, was addressed with improved techniques on photopolymers and a survey of new materials involving an electric field or dual spectrum activation.

Also the teaching of holography to future engineers was discussed during the presentation of a skillfully designed program of lectures and laboratory experimentation that expose the students to real problems occurring in industry.

In addition to the formal sessions, there were permanent demonstrations of the latest equipment in laser technology including a speckle interferometer for vibration analysis by NTE of Albuquerque, New Mexico; a 5-watt Argon Ion Laser from Coherent Laboratories, a model 1034 Industrial

Holography System including laser camera, optics and TV monitor from Newport Research Laboratories; and a New Film Camera with in situ bleaching process from Keystone Scientific Company. Other demonstrations were conducted by members of the Applied Optics Laboratories on various techniques for Nondestructive Testing.

The speakers, the chairman of the symposium, and the program committee are to be commended for the planning and the fruitful exchange of information with the participants.

We would like to acknowledge the tremendous help of the Applied Optics Laboratories of New Mexico State University, particularly Larryl Matthews, George Mulholland, and their students for arranging both the scientific and the nonscientific programs of the symposium.

We are grateful to the Newport Research Corporation, the Spectraphysics Company, the Polaroid Corporation and Keystone Scientific Company for sponsorship of the symposium through financial contribution. Finally, we are pleased to acknowledge the assistance of Miss Deborah Latshaw with the nonscientific program of the symposium.

Las Cruces, N.M.   J. J. R.
Huntsville, Ala.   J. C.

Contents

III. Industrial inspection and documentation

IV. Holographic recording materials

V. Recording equipment

VI. Teaching of holography

Contributors

John R. Andrews
Xerox Corporation
MS 0114 20D
800 Phillips Rd
Webster, NY 14580

Sergio Calixto
Centro de Investigaciones en Optica
Apartado Postal 948
Leon
Guanajuato, 37000 Mexico

H. John Caulfield
Center for Applied Optics
University of Alabama
Huntsville, AL 35899

Reinhart Damm
Xerox Corporation
MS 0114 20D
800 Phillips Rd
Webster, NY 14580

Werner E. Haas
Xerox Corporation
MS 0114 20D
800 Phillips Rd
Webster, NY 14580

Bruce Hansche
Organisation 7551
Sandia Corporation
Albuquerque, NM 87185

Lloyd Huff
Research Institute
University of Dayton
300 College Park
Dayton, OH 45469

Yasuo Kimura
Optoelectronics Research Labs
NEC Corporation
1-1 Miyazaki 4-chome
Kanagawa 213, Japan

Ole J. Løkberg
Department of Physics
Norwegian Institute of Technology
N-7034 nth
Trondheim, Norway

Larryl Matthews
Engineering Research Center
New Mexico State University
Las Cruces, NM 88003

George Mulholland
College of Engineering
New Mexico State University
Las Cruces, NM 88003

Yuzo Ono
Optoelectronics Research Labs
NEC Corporation
1-1 Miyazaki 4-chome
Miyamae-ku, Kawasaki
Kanagawa 213, Japan

Nicholas J. Phillips
Department of Physics
Loughborough University of
Technology
Loughborough, Leics
United Kingdom

Mike Rainsdon
Xerox Corporation
MS 0114 20D
800 Phillips Rd
Webster, NY 14580

Gina Sada Rightley
Department of Mechanical
Engineering
New Mexico State University
Las Cruces, NM 88003

Jean J. Robillard
College of Engineering

New Mexico State University
Las Cruces, NM 88003

Karl A. Stetson
United Technologies Research Center
MS/83
East Hartford, CT 06108

Seijin Sugama
Optoelectronics Research Labs
NEC Corporation
1-1 Miyazaki 4-chome
Miyamae-ku, Kawasaki
Kanagawa 213, Japan

Karlene Thomas
Xerox Corporation
MS 0114 20D

800 Phillips Rd
Webster, NY 14580

Vincent Toal
Department of Physics
College of Technology
Kevin Street
Dublin 8, Eire

Boyd Tuttle
Xerox Corporation
MS 0114 20D
800 Phillips Rd
Webster, NY 14580

John Watson
Kings College
University of Aberdeen
Aberdeen, UK

I
Artificial intelligence, data processing computer-generated hologram

1
The holographic basis
for intelligent machines

H. JOHN CAULFIELD

Manufacturing has come to be viewed as a concerted effort among a variety of machines, computers, and humans. In contrast, biologic manufacturing processes are far better instrumented and integrated. Indeed the word *organic* has come to mean this smooth joining of multiple functional units into a single coordinated whole. The ultimate biologic units, humans, exhibit a rather-difficult-to-define additional property called *intelligence*. The goal of the research summarized herein is the search for a hardware basis for machine intelligence.

We must begin with an understanding of the nature of intelligence. Human intelligence must understand itself sufficiently well to allow it to define its own essentials. By "essentials," I mean those aspects of intelligence that appear to be required functions as opposed to particular organic embodiment. In recent years great progress has been made in this field—a field often called *cognitive science*. My brief description of this massive progress is based loosely on the work of Pylyshyn (1).

We can think of intelligence as the joint operation of three levels of activity. At the lowest level, we have purely mechanistic hardware (or in the human case "wetware"). The behavior of that hardware can be described from the point of view of construction. If we know the inputs, we can, in principle, predict the outputs. Its behavior does not need such cognitive terms as motive, intention, emotion, or belief for explanation. It is not "cognitively penetrable." Great progress in modeling the hardware of the human brain has led to a recent explosion of activity in constructing nonbiologic hardware based on these models. These hardware have come to be called *artificial neural systems* or *neural networks*.

The next level in intelligence is a transducer of external stimuli and internal states into suitable and useful signals to serve as inputs to the hardware. Some of these transducer functions are cognitively penetrable, such as low level vision and low level hearing. Other transducers, such as high level vision, depend on expectations, beliefs, emotions, and are cognitively penetrable. Intention, goals, beliefs, and so on affect the results and are, themselves, represented as inputs to the hardware.

1.   a.   Possessing intelligence, b. Guided or directed by intellect.

2.   a.   Having or indicating a high or satisfactory degree of intelligence and mental capacity, b. Revealing or reflecting good judgement or sound thought.

3.   Able to perform some of the functions of a computer.

Figure 1.1. Webster's new collegiate dictionary (1979) definition of "intelligent."

The third layer is that of syntax and logic. It describes how ideas, concepts, and beliefs interact and are interpreted.

Our goal here is to show how holography facilitates the construction of the lowest level, the hardware basis for intelligence. It does not follow that any use, however complicated, of that hardware constitutes intelligence. Most, if not all, currently proposed uses of neural nets work entirely at the lowest level. They are not cognitively penetrable and not deserving of the word intelligence. On the other hand, intelligence can never emerge without an underlying level of hardware of sufficient complexity.

This brief discussion is intended to set the context for the work reported. It is also intended as a disclaimer of the type of confusion represented in Figure 1.1. This confusion appears often in the technical literature, albeit in less crude form.

Neural networks

A very simple description of neural networks follows. The intent is to state the problem to be solved, not to show the usefulness of neural networks. That information is readily available (2).

Let us consider a single "neuron" as shown in Figure 1.2. The neuron receives real inputs I_1, I_2, ..., I_N from many other neurons. These inputs are summed algebraically,

$$I = \sum_i I_i \qquad (1.1)$$

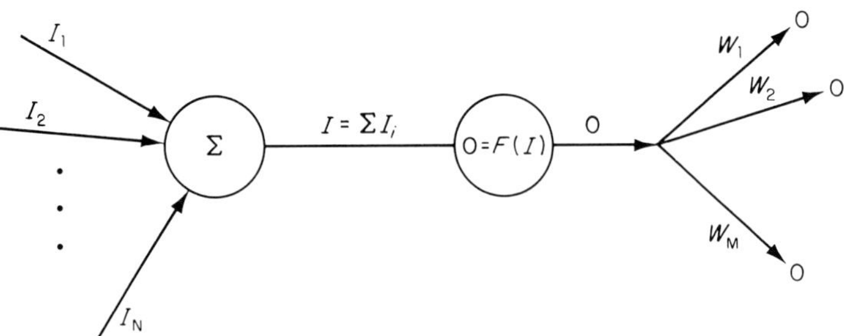

Figure 1.2. A simplified neuron.

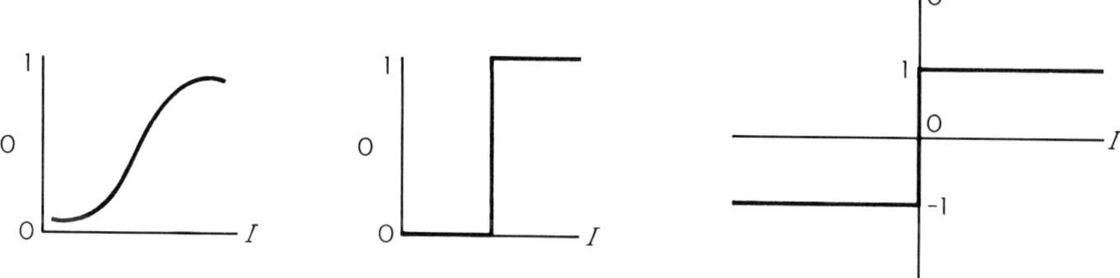

Figure 1.3. Three common forms of I to 0 transformations used in neural networks.

There is an operator that generates an output signal 0. Often this operator is one of the forms shown in Figure 1.3. The 0 signal is then distributed to M other neurons with real weights W_1, W_2, ..., W_M. Usually neural networks are arranged in multiple layers. In some, all information transfer is undirectional (feed forward). In general, feed forward, feed backward, and feed sideways (e.g., lateral inhibition) are used.

Most work in using neural networks involved supervised learning. Input–output datasets (vectors) are used to adjust the weights to achieve satisfactory performance. The pattern of weights is sometimes called *long-term memory*. The act of insertion of an input signal vector and extraction of an output signal vector is called *short-term memory*. In human intelligence short-term memory and long-term memory occur in the same hardware (wetware). Indeed, short-term memory affects long-term memory.

I argue that this cojoining is biologically understandable but not a defining characteristic of a neural network. In the scheme suggested below, totally different hardware is used for this type of memory.

Need for massive interconnect

In the machinery of human intelligence, the numbers of neurons and interconnections are well beyond our wildest dreams of technical duplication. The brain and sensor processors of humans tend to involve "clumps" of neurons densely interconnected. That is, within such clumps virtually every neuron is connected to virtually all others. These clumps are far less densely interconnected with others. The numbers of neurons in a densely interconnected portion of the brain may be up to a few million. If we take this as a good guess as to the maximum number of neurons that need to be densely interconnected in an intelligent system, we can ask: "What technical approaches are conceivable for connecting each of 10^6 'neurons' to each of another 10^6 'neurons'?" How many interconnections is this? The answer is: 10^{12}. Furthermore, for speed we need to make these interconnections in parallel. Time sequential interconnect will not do. If we could make a full 10^6 interconnections each millisecond (an outrageous but not impossible task), it would take "only" 10^3 seconds or 17 minutes for flow of signals between the two sets of neurons.

Interconnection technologies

Biology can do it. Our brains are proof that 10^{12} parallel interconnections can be made in biology with low volume and fast response times. The only trouble is, we have not learned how to construct and teach biologic brains other than nature's way: the procreation of new humans. What we seek is a technologically controllable, predictable neural network. Human brains meet neither criterion. In the next century it may be possible to engineer wetware. Until then, biology cannot be our solution.

Electronics cannot do it. I understand that impossibility is a strong word, so let me offer a constructive reduction to absurdity. As we push toward submicron charge carriers, it is clear that submicron wires with insulators bringing their diameters to 1 μm are not impossible. If we could pack these on 1 μm centers, 10^{12} of them could fit into an area of only 1 square meter. We will totally ignore such improbabilities as ways to adjust 10^{12} weights, ways to "scramble" the wires to make the right inputs go to the right

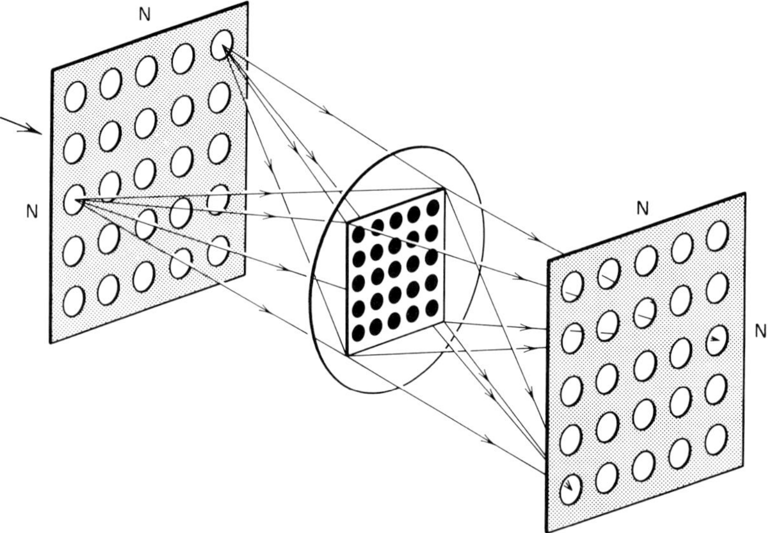

Figure 1.4. Optical neural network. An $N \times N$ hologram array is imaged onto an $N \times N$ output array through an SLM. Each hologram illuminates the SLM with a unique pattern of light. All work in parallel. The SLM contains input information. The short term memory traces of this neural appear in parallel on the right. A lens near the SLM does the imaging, which is really phase conjugation of recording beams.

outputs, the immense volume and weight of such an interconnect array, capacitive coupling effects, and so on. Let us assume that all of those problems are solved. Now all we have to do is wire the interconnection array to the two layers of neurons. This requires only 4×10^{12} welding operations on submicron wires spaced 1 μm apart. Let us assume that this can be done and can even be done with the literally incredible speed of 4000 per second. This would take only 10^9 seconds or about 30,000 years. My assertion of impossibility is sociologic. We would surely lose interest in the project before we could complete it.

What biology cannot do controllably and electronics cannot do at all, optics can do today. Consider the arrangement show in Figure 1.4. The data are input into an $N \times N$ spatial light modulator (SLM). An $N \times N$ array of holograms illuminated simultaneously produce the interconnections. Let the transmission of the k, l element of the SLM be a_{kl}. Let the light intensity produced by the i, j hologram at the k, l position on the SLM be T_{ijkl}. A lens images the hologram array into the output array (the image of the hologram array). Let the i, j element of the output array be b_{ij}. Clearly

$$b_{ij} = \sum_{k,l} T_{ijkl} a_{kl}. \tag{1.2}$$

some readers will recognize T_{ijkl} as a four-dimensional tensor. If i, j, k, and l run from 1 to N, T_{ijkl} has N^4 components, for $N = 10^3$, $N^4 = 10^{12}$.

Partitioning the two-memory functions

My scheme is to train the long-term memory electronically. Neural network equations are easy to partition and stitch, therefore no electronic computer needs to handle 10^{12} interconnections at any instant. On the other hand, electronic computers have the flexibility and accuracy to implement the subtle learning rules neural network theorists derive. Learning may take a long time even on a supercomputer. This is not surprising in view of the unprecedented complexity of the hardware. Once the learning is complete, it may take a few days to write the hologram(s). Copying the hologram, however, is far quicker—say, a few minutes. Finally, using hardware for short-term memory will require roughly a millisecond (SLM cycle time limited).

Reprise

The lowest of the three levels of an intelligent technology appears to be feasible using holographic technology. Furthermore, this combination of very high speed and very high complexity in engineered hardware ($\sim 10^{15}$ connections/second) is well beyond the current state of the art in biology and essentially impossible electronically. To my knowledge, this is the first demonstration that there is an important computing function that is impossible for electronics and currently feasible with optics.

Acknowledgment

This work was done for Office of Naval Research under Contract Number N00014-86-K-0591.

References

1. Pylyshyn, Z. W. *Computation and Cognition, Toward a Foundation for Cognitive Science.* MIT University Press (1984).
2. Caulfield, H. J. *Parallel N^4 Weighted Optical Interconnections, Applied Optics* **26,** 4039 (Oct. 1, 1987).

2
The foundations of holography

NICHOLAS J. PHILLIPS

The science of holography has developed along lines that are very much dictated by the availability of suitable recording materials. Thus, silver halides were used in Lippmann's era (c 1891) for the recording of interference patterns created by the interference of spatially incoherent light with its own reflection. Because holography really accelerated, after the invention of the laser in 1960, several key materials have helped the progress of the subject. Probably the most significant of these would now be seen as the silver halides, dichromated gelatin, photoresist, and in various guises the photopolymers. That other media exist is understood but what is required is a combination of practical usage and suitable and repeatable process techniques. The recording process is complex and may involve both a latentisation stage followed by some sort of developmental procedure. With some materials, latentisation as in the silver halides is apparently absent, although a more careful look usually reveals some similarities. We shall examine some theoretical modeling of the image-forming process and offer some conclusions based on a simplified approach. A key problem is the question of whether the optical pattern that is recorded is accurately replayed by interrogation of the recording material. Although this question has been addressed in terms of, for example, the linearity of the recording process, the fundamental issue would appear to be whether the location of an image point, as replayed by the hologram, is in exact correspondence with the object point from which it originates. Most holographic images are used in applications of sufficiently low precision not to warrant an answer. We shall address here the modeling of a particular type of image-forming mechanism, namely diffusion transfer in a silver halide medium. The generalization of the results to other media is discussed and comparisons are attempted.

Theoretical modeling of an image-forming process

Let us now look at a model of the recording process in a silver halide medium. The grain structure of the medium is regarded as of secondary importance as compared with the diffusion transfer mechanism proposed for the image modulation process. Central to the issue is the point that for a holographic layer to exhibit ultimate precision in the imaging process is the retention of the original molecular population of the layer. Any removal of or addition to the original molecular population must perforce lead to a shrinkage or swelling of the layer. With silver halide media it has transpired that methods involving fixing of the layer or solvent bleaching resulting in removal of silver have given way to methods of diffusion transfer that effectively keep the layer thickness more or less at its initial value. The reason why such transfer methods are now seen to be of paramount importance is that other approaches have led to shrinkage and as a result a distortion of the interference structure in the layer.

Let us begin by considering a one-dimensional model of an optical pattern in which the exposure process followed by a developmental procedure has led to the formation of a spatial pattern of silver atoms, thus,

$$n_{Ag0}(x) = \varphi(x) \tag{2.1}$$

These atoms will have been formed from an initial population of silver bromide molecules of number density $n_{00} = $ constant. Note that we ignore the grainy structure of a real medium but such an approximation does not impede an understanding of the physics. Evidently the initial population of silver bromide molecules is given by

$$n_{AgBr0} = n_{00} - n_{Ag0}(x) \tag{2.2}$$

Now consider the effect of the chemical attack of a rehalogenating bleach. Such a bleach plays two roles as it must first ionize the silver atoms and then cause their precipitation by interaction with bromide ions present in the bleach solution. That such a bleach creates image modulation can be seen as follows: silver ions will be created in a given spatial position but can then diffuse before meeting a precipitating halide ion. We can construct rate equations thus

$$\frac{\partial n_{Ag^+}}{\partial t} = \frac{n_{Ag}}{\tau_j} \tag{2.3}$$

where n_{Ag^+} is the number density of silver ions. Here τ_j is a characteristic time of ionization, which is controlled by the bleach solution in its concentration and proportions. We may also write

$$\frac{\partial n_{Ag}}{\partial t} = \frac{-n_{Ag}}{\tau_j} \tag{2.4}$$

as silver ions are created at the expense of silver atoms. We can write a complete description of the population of silver ions in the form

$$\frac{\partial n_{Ag^+}}{\partial t} = +\frac{n_{Ag}}{\tau_j} - \frac{n_{Ag^+}}{\tau_p} \frac{D\partial^2 n_{Ag^+}}{\partial x^2} \tag{2.5}$$

Here τ_p is the characteristic time before precipitation of the ions (in the form of an AgBr molecule) by the bleach and D is a diffusion coefficient characteristic of the medium and its interaction with the mobile ions. Equation 2.4 has the solution

$$n_{Ag} = n_{Ag0} \exp(-t/\tau_j) \tag{2.6}$$

and we approach the solution of Equation 2.5 by a transform method.

Let us now define the Fourier transform of a variable $q(x)$ in the form

$$\overline{q(k)} = \frac{1}{\sqrt{2\pi}} \int_{-\infty}^{\infty} q(x)e^{ikx}\, dx \tag{2.7}$$

Now we note that

$$\overline{\frac{\partial^2 q}{\partial x^2}} = \frac{1}{\sqrt{2\pi}} \int_{-\infty}^{\infty} \frac{\partial^2 q}{\partial x^2} e^{ikx}\, dx = \frac{1}{\sqrt{2\pi}} \left\{ \left[e^{ikx} \frac{\partial q}{\partial x} \right] - ik \int_{-\infty}^{\infty} \frac{\partial q}{\partial x} e^{ikx}\, dx \right\}$$

$$= \frac{1}{\sqrt{2\pi}} \left\{ \left[e^{ikx} \frac{\partial q}{\partial x} \right] - ik \left([e^{ikx}q] - ik \int_{-\infty}^{\infty} q e^{ikx}\, dx \right) \right\} \tag{2.8}$$

If we can assume that both q and its first derivative vanish at $\pm\infty$, perhaps encouraged by a Gaussian beam profile, then it follows that

$$\overline{\frac{\partial^2 q}{\partial x^2}} = -k^2 \bar{q} \tag{2.9}$$

Equation 2.5 can now be written in the form

$$\frac{\partial \overline{n_{Ag^+}}}{\partial t} = +\frac{\overline{n_{Ag}}}{\tau_j} - \frac{\overline{n_{Ag^+}}}{\tau_p} - k^2 D \overline{n_{Ag^+}} \tag{2.10}$$

But from Equation 2.6, this can be written

$$\frac{\partial \overline{n_{Ag^+}}}{\partial t} = \frac{\overline{n_{Ag0}}}{\tau_j} \exp\left(-\frac{t}{\tau_j} \right) - \frac{\overline{n_{Ag^+}}}{\tau_p} - k^2 D \overline{n_{Ag^+}} \tag{2.11}$$

This must now be solved for $\overline{n_{Ag^+}}$ as a function of t. This is easily shown to have the solution

$$\overline{n_{Ag^+}} = \frac{\overline{n_{Ag0}}}{\tau_j} \left\{ \exp\left(-\frac{t}{\tau_j} \right) - \exp\left[-t\left(\frac{1}{\tau_p} + k^2 D \right) \right] \right\} \bigg/ \left(\frac{1}{\tau_p} + k^2 D - \frac{1}{\tau_j} \right) \tag{2.12}$$

Now we note that the precipitated bromide is n_{AgBr} and is given by

$$\frac{\partial n_{AgBr}}{\partial t} = +\frac{n_{Ag^+}}{\tau_p} \tag{2.13}$$

or in its transformed form, we have

$$\frac{\partial \overline{n_{AgBr}}}{\partial t} = \frac{\overline{n_{Ag^+}}}{\tau_p} \tag{2.14}$$

therefore, as $n_{AgBr} = 0$ at $t = 0$, i.e., there has been no precipitation of AgBr by the bleach at this time

$$n_{AgBr} = \frac{\dfrac{n_{Ag0}}{\tau_p \tau_j}\left[-\tau_j(\exp(-t/\tau_j) - 1) + \dfrac{1}{\left(\dfrac{1}{\tau_p} + k^2 D\right)}[\exp(-t(1/\tau_p + k^2 D)) - 1]\right]}{\dfrac{1}{\tau_p} + k^2 D - \dfrac{1}{\tau_j}}$$

$$(2.15)$$

In the limit $t \to \infty$ all the silver is used up and we have

$$\overline{n_{AgBr}} \to \frac{\dfrac{\overline{n_{Ag0}}}{\tau_p \tau_j}}{\dfrac{1}{\tau_p} + k^2 D - \dfrac{1}{\tau_i}}\left[\tau_j - \dfrac{1}{\left(\dfrac{1}{\tau_p} + k^2 D\right)}\right] = \frac{\overline{n_{Ag0}}}{(1 + k^2 D \tau_p)} \qquad (2.16)$$

Using the inversion theorem, we have the precipitated halide in the form

$$n_{AgBr} = \frac{1}{\sqrt{2\pi}} \int_{-\infty}^{\infty} \frac{\overline{n_{Ag0}}(k) e^{-ikx}}{1 + k^2 D \tau_p} \, dk \qquad (2.17)$$

We can further simplify the result using the convolution theorem

$$\int_{-\infty}^{\infty} F(k) G(k) e^{-ixk} \, dk = \int_{-\infty}^{\infty} g(\eta) f(x - \eta) \, d\eta \qquad (2.18)$$

where F and G are the Fourier transforms of f and g, respectively. To revise the structure of Equation 2.17 we need to identify the appropriate functional forms. Thus, we note that

$$\int_{-\infty}^{\infty} e^{-\alpha|x| + ikx} \, dx = \frac{2\alpha}{\alpha^2 + k^2} \qquad (2.19)$$

we then choose $\alpha^2 = 1/D\tau_p$ and it follows that

$$n_{AgBr} = \frac{\alpha}{2} \int_{-\infty}^{\infty} \overline{n_{Ag0}}(k) G(k) e^{-ikx} \, dx \qquad (2.20)$$

where $G(k)$ is the Fourier transform of $e^{-\alpha|x|}$. Thus, the convolution theorem reveals the result

$$n_{AgBr} = \frac{\alpha}{2} \int_{-\infty}^{\infty} e^{-\alpha|\eta|} n_{Ag0}(x - \eta) \, d\eta \qquad (2.21)$$

which completes the solution. From this result, we can calculate the final distribution of the precipitated silver halide molecules. The question is whether this result reveals a correspondence of the positional distribution of molecules with the antinodal pattern of the recording light.

Special cases

The most simple case is that of a sinusoidal distribution of developed silver. Thus, for example,

$$n_{\mathrm{Ag0}} = \frac{n_0}{2}\left[1 + \cos\left(\frac{2\pi x}{d}\right)\right] \tag{2.22}$$

Here d is the pattern wavelength and n_0 is the peak value of the silver atom density. From Equation 2.22, we have

$$n_{\mathrm{AgBr}} = \frac{\alpha}{2}\int_{-\infty}^{\infty} e^{-\alpha\eta}\frac{n_0}{2}\left[1 + \cos\left(\frac{2\pi(x-\eta)}{d}\right) + 1 + \cos\left(\frac{2\pi(x+\eta)}{d}\right)\right]d\eta \tag{2.23}$$

$$= \frac{n_0}{4}\left[2 + \frac{2\alpha^2\cos\left(\dfrac{2\pi x}{d}\right)}{\alpha^2 + \dfrac{4\pi^2}{d^2}}\right] \tag{2.24}$$

Now we calculate the total distribution of the silver halide atoms. Thus, we remember that

$$n_{00} = n_{\mathrm{AgBr0}} + n_{\mathrm{Ag0}} \tag{2.25}$$

that is, there are either atoms of silver or silver bromide molecules present in any one locality. The total halide population density is finally

$$n_{\mathrm{AgBr0}} + n_{\mathrm{AgBr}}$$

$$= n_{00} - \frac{n_0}{2}\left[1 + \cos\left(\frac{2\pi x}{d}\right)\right] + \frac{n_0}{2}\left[1 + \frac{\alpha^2}{\alpha^2 + \dfrac{4\pi^2}{d^2}}\cos\left(\frac{2\pi x}{d}\right)\right] \tag{2.26}$$

$$= n_{00} - \frac{n_0}{2}\cos\left(\frac{2\pi x}{d}\right)\frac{\dfrac{4\pi^2}{d^2}}{\alpha^2 + \dfrac{4\pi^2}{d^2}} \tag{2.27}$$

Comparing this distribution with that of the original silver, Equation 2.22, we note both a modification of contrast and spatial phase of the pattern. This then is the problem that we have addressed. It appears that the spatial phase of original interference pattern and final distribution of atoms do not correspond.

Clearly, if the pattern of silver is given by Equation 2.22, then it matters little whether the phase is shifted or not. Such a distribution corresponds to a simple plane grating approximation—no information is carried by the recording.

Suppose now that we record the interference of a plane wave with the wave fronts from a point source S as follows (Fig. 2.1):

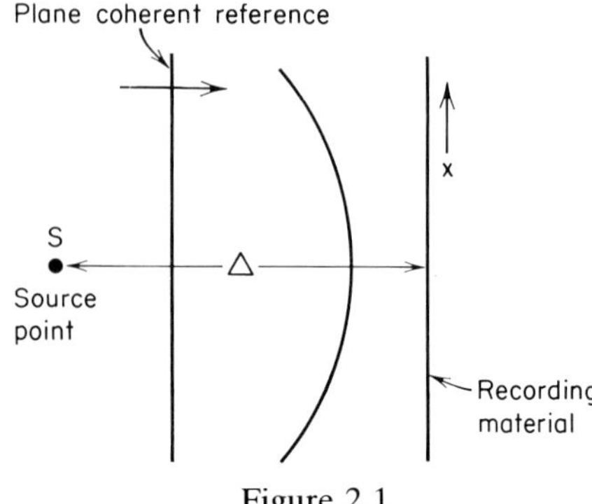

Figure 2.1

The total wave amplitude can be represented by

$$A_R + A_S \exp\left(\frac{ikx^2}{2\Delta}\right) \tag{2.28}$$

where $k = 2\pi/\lambda$, A_R is the reference amplitude, and A_S the signal amplitude. We might reasonably expect a silver pattern of the form

$$n_{Ag0} = \frac{n_0}{2}\left[1 + \cos\left(\frac{2\pi x^2}{\lambda\Delta}\right)\right] \tag{2.29}$$

The quadratic dependence on x must now be accounted for in the integration of Equation 2.22.

We now have

$$n_{AgBr} = \frac{\alpha}{2}\int_0^\infty e^{-\alpha\eta}\frac{n_0}{2}\left[1 + \cos\left(\frac{2\pi(x-\eta)^2}{\lambda\Delta}\right) + 1 + \cos\left(\frac{2\pi(x+\eta)^2}{\lambda\Delta}\right)\right]d\eta \tag{2.30}$$

$$= \frac{n_0}{2} + \frac{\alpha n_0}{4}\int_0^\infty e^{-\alpha\eta}\left[\cos\left(\frac{2\pi(x-\eta)^2}{\lambda\Delta}\right) + \cos\left(\frac{2\pi(x+\eta)^2}{\lambda\Delta}\right)\right]d\eta \tag{2.31}$$

The integrals involved in Equation 2.31, as yet, can only be calculated by numerical methods, but fortunately we can approximate to the solution in the limit of high levels of diffusion $D \to \infty$. We note that

$$\alpha^2 = \frac{1}{D\tau_p}$$

and hence that $\alpha \to 0$ as $D \to \infty$. In this limit, the modulated terms in the precipitated halide disappear and we have

$$n_{\mathrm{AgBr}} \rightarrow \frac{n_0}{2} \qquad (2.32)$$

Thus, the total concentration of silver bromide is finally

$$(n_{\mathrm{AgBr}})_{\mathrm{tot}} = n_{\mathrm{AgBr0}} = n_{\mathrm{AgBr}} = n_{00} - n_{\mathrm{Ag0}} + n_{\mathrm{AgBr}}$$

$$= n_{00} - \frac{n_0}{2}\left[1 + \cos\left(\frac{2\pi x^2}{\lambda \Delta}\right)\right] + \frac{n_0}{2} \qquad (2.33)$$

We note that as $x \rightarrow \infty$ then $(n_{\mathrm{AgBr}})_{\mathrm{tot}} \rightarrow n_{00} - n_0/2$. We also note that if we let $2\pi x^2/\lambda D \rightarrow \pi$ then $(n_{\mathrm{AgBr}})_{\mathrm{tot}} \rightarrow n_{00} + n_0/2$ so that the pattern center exhibits a *minimum* in the level of the final halide concentration.

Comparison of results between the silver halides and DuPont's photopolymer

It is now of interest to compare two apparently widely discrepant media on the basis of the preceding modeling. Surprisingly, both materials can be tackled with the same physical approach. The DuPont polymer is thought to behave according to the Colburn and Haines model. The basic idea is that the first phase of exposure causes polymerization of the monomer in the locality of the antinodes of the pattern. The remaining monomer then diffuses to fill the spaces in concentration left by its local polymerization. Of course, this is a complex process as D may well be affected by the polymerization that has gone before. With the assumption that D is constant and large enough, then the same approximation of $D \rightarrow \infty$ can be invoked but the differences are now profound. It must be remembered that it is molecular concentration and related polarizability factors that cause the necessary phase shifts in the hologram. Thus, in any locality it is the sum

$$\sum_i n_i \alpha_i$$

that controls the local optical path in the recording layer. Here n_i is the concentration of the *i*th species of molecule and α_i its polarizability.

Now suppose that the optical pattern exhibits a maximum of intensity at the center of the interference pattern denoted in Equation 2.29. Clearly this will lead to a maximum in initial polymer generation at the pattern center and the monomer will move to fill the depression of its concentration. It is then quite clear that the phase shift induced at the pattern center will be a *maximum* not a *minimum* as in the silver halide case. This leads to the conclusion that an indeterminacy of position of the source S may arise in the recording process. Let us now adjust the phase of the coherent reference and write Equation 2.28 in the form

$$A_0 e^{i\phi} + A_s \exp\left(\frac{ikx^2}{2\Delta}\right) \tag{2.34}$$

Adjust A_0 to equal A_s and the pattern intensity becomes

$$I = A^2\left[e^{i\phi} + \exp\left(\frac{-ikx^2}{2\Delta}\right)\right]\left[e^{-i\phi} + \exp\left(\frac{+ikx^2}{2\Delta}\right)\right] \tag{2.35}$$

$$= A^2\left[2 + 2\cos\left(\frac{kx^2}{2\Delta} + \phi\right)\right] \tag{2.36}$$

Here the pattern center is adjustable from dark to bright, merely by altering ϕ. We might address this adjustment differently by altering the distance Δ by an increment δ. The distance factor r in the diffraction problem becomes

$$r^2 = \Delta^2 + x^2 \rightarrow (\Delta + \delta)^2 + x^2 \tag{2.36}$$

so that

$$r \approx \Delta + \delta + \frac{x^2}{2\Delta^2} \tag{2.37}$$

In the limit of $x \rightarrow 0$ then $r \approx \Delta + \delta$. We merely have to choose δ appropriately to change the center of the pattern from a bright to a dark. Evidently, an increment $k\delta = \pi$ is sufficient to change the pattern over its complete range.

The interpretation of the recording is now seen to have an ambiguity associated with the depletion of refractive material in the case of the silver halides and its enhancement in the case of the polymer.

Consideration of ultimate precision in holograms

The materials modulated by diffusion are clearly desirable and necessary for flatness and absolute freedom from shrinkage of the recording layer but they carry with them an apparent penalty of ambiguity of location of the point image in the cases cited. It looks simplistic, as if the polymer has advantages over the halide material. Clearly, those methods of recording that use programmed removal of molecules such as e beam into resist are different in that they perform their task without addressing this problem.

Conclusions

Familiarity with holograms can be accompanied by ignorance of fundamental issues. The hologram may be used to record very precise information. For example, it is finding use in areas of precision microlithography.

This chapter seeks to point out that the ultimate precision of the image may be limited by the image-forming mechanism. Information theory would suggest that the act of recording must perforce degrade the information content of what is recorded. That obvious differences occur between two quite unrelated media is not surprising. That such a similarity of physical description can be invoked is fascinating.

3

Holographic stereograms of computer-generated objects made using a liquid crystal spatial light modulator

JOHN R. ANDREWS, BOYD TUTTLE,
MIKE RAINSDON, REINHART DAMM,
KARLENE THOMAS, and WERNER E. HAAS

Holographic stereograms are an important way of generating three-dimensional images from photographs of real objects (1, 2) or of computer-generated three-dimensional objects (3). The two-dimensional views from which the stereogram is synthesized are usually derived from a series of photographs of a real scene taken from the correct perspective location (4) or from pictures of the computer graphics screen (or film recorder) displaying the correct perspectives of the computer-generated object (3). In both cases the photographic intermediate is incoherently illuminated and photographed. Replay in the holographic recording step is done with coherent illumination through an appropriate projection system.

The photographic intermediate in going from the series of two-dimensional perspectives to a stereogram is inconvenient for a number of reasons. The expense of the specialized (pin-registered) camera and film transport equipment requires substantial capital and rather bulky optical arrangements. The time delay and expense related to the pictorial film processing can lengthen the turnaround time for the process by hours or even days. The entire process requires a lot of human intervention for the film handling is not easily amenable to automation.

The search for a rapid turnaround and a process that can be automated lead us to explore the use of two-dimensional spatial light modulators to replace the film in holographic stereogram production. To our knowledge, this is the first report of the realization of holographic stereogram production from computer-generated images using a spatial light modulator. Spatial light modulators have previously been used in optical information processing (5–9).

The generation of the holographic stereograms followed a sequential

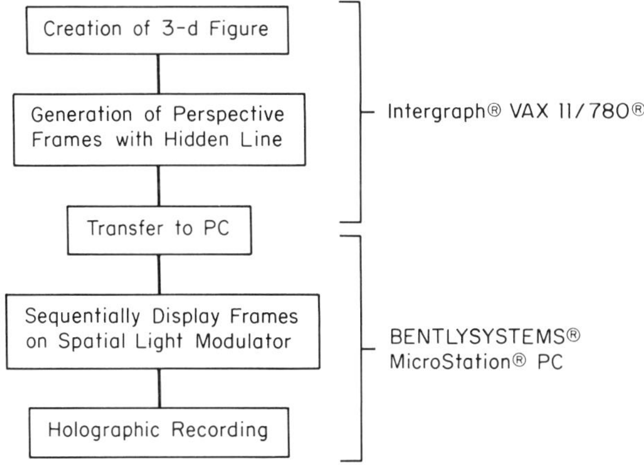

Figure 3.1. The generation of the holographic stereograms.

pattern of development as shown in Figure 3.1. The original images were created on the INTERGRAPH® computer-aided design (CAD) system running on a VAX® 11/780. The original three-dimensional design files were used in a program that generated a series of 10 to 100 projection views in appropriate angular increments (1°–5°) using a hidden line elimination format (HLine). Computation times varied from 1–6 hours. The projection views were then transferred via an RS-232 port or floppy disk to a personal computer-based system for display in the holographic step. The personal computer software for displaying the images on the display panel was the BENTLEYSYSTEMS® MicroStation®. We used a WELLS-AMERICAN® personal computer having a color graphics adapter (CGA). For display of the computer-generated images we used a KODAK DATASHOW® liquid crystal display panel having a resolution of 640 × 200 pixels and a display size of 20.3 × 15.2 cm. This unit was chosen because of its compatibility with the CGA driver on the personal computer.

The optical arrangement for making the holographic stereograms is shown in Figure 3.2. The arrangement, with the exception of the spatial light modulator, is similar to conventional arrangements (3). A single mode argon ion laser is used as the coherent light source. This is split into reference and object beams, each of which is expanded into a plane wave through a telescope. The reference beam directly illuminates the holo graphic plate, whereas the object beam passes through the liquid crystal spatial light modulator and onto a diffusing screen. Some of the diffused light from the image projected onto the screen strikes the holographic plate. A moveable slit localizes each exposure with the object and reference beams to a narrow band. The recorded slit holograms are in a horizontal sequence that corresponds to the sequence of perspectives produced with the hidden line calculation. The entire experiment is under the control of a

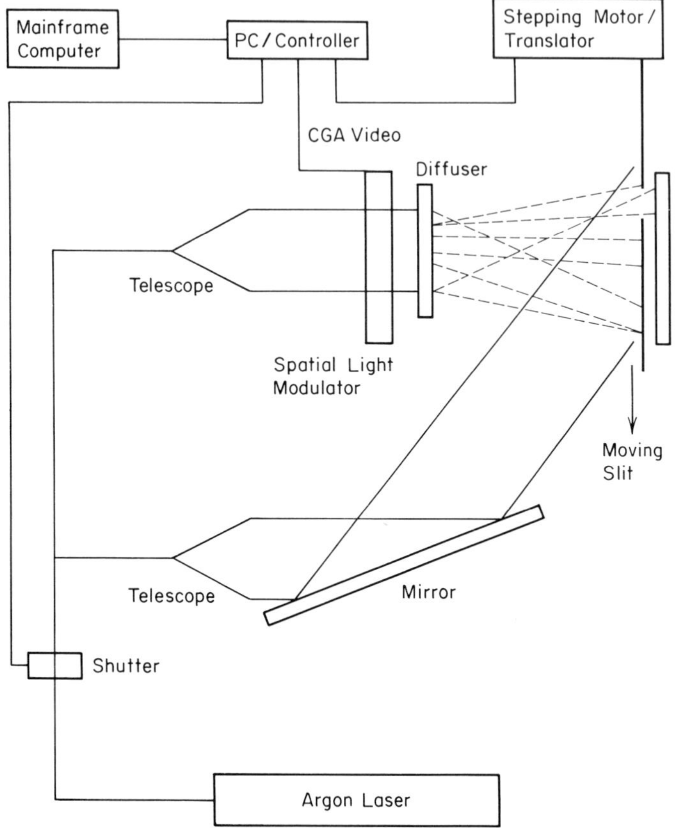

Figure 3.2. The optical arrangement for making the holographic stereograms.

personal computer. It puts an image on the liquid crystal display, opens the shutter for the laser exposure, advances the slit, and puts a new image on the display. This cycle is followed until all of the perspective images in the file have been recorded as one of the holographic slit images. A typical stereogram consisted of 30 frames in $2°$ increments with exposure times of 0.1 to 1 seconds per frame on 20×25 cm holographic plates. The hologram, when developed and bleached, is laser viewable. A white light viewable stereogram can be made by making an image plane copy (3, 10). A single step technique for generating white light viewable holograms is the single step called Cross method (11, 12).

Photographs of a laser viewable holographic stereogram made according to the preceding procedure are shown in Figure 3.3 for images seen from two perspectives, one at the far right and the other at the far left. The holographic stereogram produced by this method offers horizontal parallax and the visual perception of three dimensions as is the case where photographic intermediates are used. The contrast is only 2 for this device.

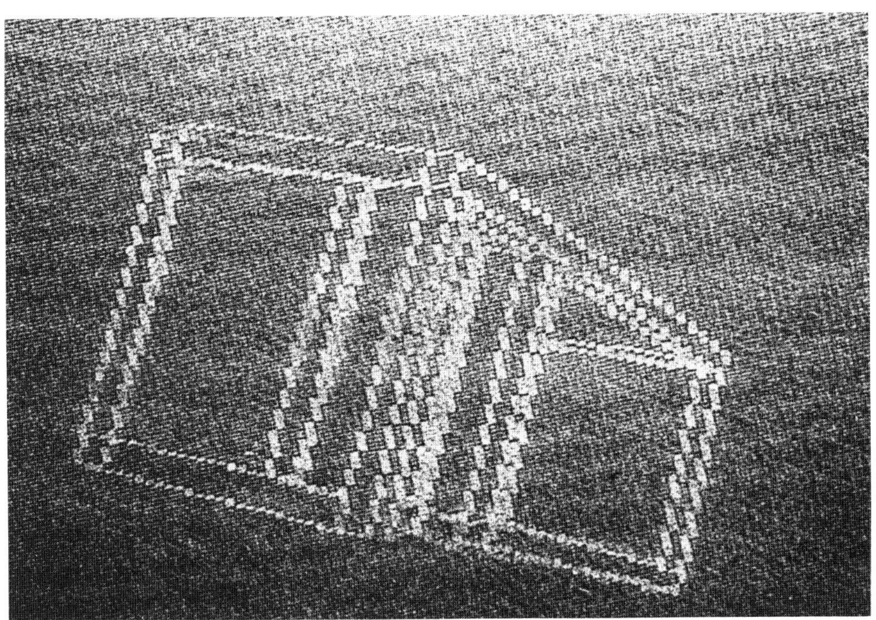

Figure 3.3. Photograph of holographic stereogram of computer-generated object.

This means that the entire diffusing screen is partially illuminated during each exposure. The result in the final image is that the object appears to reside in a cloud. Close examination of the photographs of the hologram reveals several features of the hologram. The "staircase" diagonal lines are due to the limited resolution of the CGA graphics and appear the same on the CRT monitor. The 640×200 pixel format supporting only the CGA video is substantially less than what is available from the software. The limited resolution restricts the amount of detail that can be reproduced. Low spatial frequency fringes can be seen in the diffuse background, which are due to interferences between the glass–air interfaces in the liquid crystal display panel. A fine grid pattern are the result of shadows cast by the pixel leads in the display.

In spite of the current shortcomings in the spatial light modulator display technology, the technique described here provides a quick and automated means of converting computer-generated three-dimensional objects into autostereoscopic three-dimensional presentations of the object. Another commonly available two-dimensional spatial light modulator, the Realistic® model 16-153 liquid crystal television set, has a contrast of 20 although the number of pixels is small and the signal format is not readily compatible with computer monitors. The rapid development in the spatial light modulator technology for optical processing will lead to substantially enhanced performance both in resolution and contrast. Improved spatial light modulators and the single step production of white-light viewable holographic stereograms present the possibility for rapid and fully autom- ated production of autostereoscopic three-dimensional representations of computer-generated objects.

Acknowledgment

The authors would like to thank John Landigan of Hahn Graphics for loan of the KODAK DATASHOW.

References

1. Redman, J. D. *Proc. SPIE* (1968) 117.
2. McCrickerd, J. T., and George, N. *Appl. Phys. Lett.* **12,** 10 (1968).
3. King, M. C., Noll, A. M., and Berry, D. H. *Appl. Opt.* **9,** 471 (1970).
4. George, N., McCrickerd, J. T. and Chang, M. M. T. *Proc. SPIE* (1968) 161.
5. Bleha, W. P., Lipton, L. T., Wiener-Avnear, E., Grinberg, J., Reif, P. G., Casasent, D., Brown, H. B., and Markevitch, B. V. *Opt. Eng.* **17,** 371 (1978).
6. Warde, C., Weiss, A. M., Fisher, A. D., and Thackara, J. I. *Appl. Opt.* **20,** 2066 (1981).
7. Liu, H.-K., Kung, S. Y., and Davis, J. A. *Opt. Eng.* **25,** 853 (1986).

8. Young, M. *Appl. Opt.* **25,** 1024 (1986).
9. Yu, F. T. S., Jutamulia, S., and Huang, X. L. *Appl. Opt.* **25,** 3324 (1985).
10. Benton, S. A. *Proc. SPIE* **391,** 2 (1983).
11. Cross, L. Multiplex Holography. SPIE Annual Meeting, San Diego, CA, August 1977.
12. Huff, L., and Fusek, R. L. *Proc. SPIE* **215,** 32 (1980).

4
The principles and applications of electronic speckle pattern interferometer

OLE J. LØKBERG

Looking back into the history of technology we find many examples of the time being ripe for an invention. The introduction of electronic speckle pattern interferometer (ESPI) (also called TV-holography) may serve as a typical example of this principle. In this case the main impetus came from the frustration of the holographic worker over the slow, cumbersome recording and developing process of conventional holographic film. It also became apparent that for many technical applications and measurements, the three-dimensional display was not of importance. Within 1 year four independent groups had reported on the use of the TV-system for recordings of holograms (1–4). Interesting enough, these pioneers semed to have reached their solutions from different viewpoints. Macovski and colleagues (1) at Stanford, worked on a video system for transmission of holograms and used the system for interferometric purposes as a natural extension. The English group in Loughborough (2) headed by Butters and Leendertz (2) preferred the viewpoint of speckle correlation to develop and explain the technique. Finally, Schwomma (3) from the Eumig company in Austria and Kopf (4) from the Federal Republic of Germany considered the technique to be direct recordings of primitive holograms for interferometric purposes.

After the initial flurry of activity that always follows the introduction of a new technique, the research interest almost waned (apart from a few fanatical groups) and the technique never made the expected progress within industrial applications.

At this point, let us discuss briefly why the promised land did not appear? Many answers to this question have been put forward. It has been suggested that the quality of the early recordings were inferior to what we get today, but this is simply not true. The fringe patterns produced a decade ago by the Eumig system (3), which used a high-resolution camera and an image memory tube, had an excellent resolution with very fine speckles. The only recent significant gain in terms of pictorial quality is due to speckle averaging introduced by Slettemoen (5), but this technique works only for documentation of periodic movements. The answer to the initial question is complex. Among optical scientists there was a feeling that the optical part of the system was already complete (which is perhaps true as the only contribution the last 10 years is again due to Slettemoen (5) with his speckle reference system). Some of the holography people might have been kept away from the combination of optics and video electronics, which is more complex and expensive than holography. Most important, however, few, if any, working within hologram interferometry realized the full potential of the Electron Speckle Pattern Interferometer (ESPI) technique in terms of stability, measurement of range, and ease of handling. As for the industrial lack of interest, I feel that the industry simply was not ready for the technique just as hologram interferometry had a hard time being accepted as a measuring tool in those days. In addition the commercial systems produced in those days were heavy and difficult in use.

Today we witness a renewed interest in the technique due to several reasons. First, it is obvious that for industrial on-line inspection even the thermoplastic hologram interferometry system is too slow and inflexible and the speed of video systems are needed. Second, the rapid emergence of digital video techniques, computer hardware and software make it possible to realize very easily many of the ideas suggested and painfully tested out by analog techniques years ago.

In this chapter, we review the basics of the technique and its present stage of development and cite some examples of measurements performed in our group. We also attempt to peer into the crystal ball and predict future areas of development.

The basics of electronic speckle pattern interferometer

The ESPI principle can be explained using the speckle theory, or comparing to Moire effects, or we can use a holographic analogy. For the most commonly used ESPI system, where a uniform reference wave is used, the holographic analogy is obvious and helps to understand the technique.

The opto-mechanical construction of an ESPI system is very similar to conventional hologram interferometry set-ups and we comment only briefly on the various blocks of the ESPI flow-diagram depicted on Figure 4.1. Readers who are interested in a more detailed description should consult the literature (6, 7).

As light source, the commonly used lasers, such as He–Ne, Ar, and Rb, are used. Note that the recording medium, the TV-target, usually has a far better response in the infrared than in holographic films, which opens for a broader choice of lasers than in hologram interferometry. For example, the 1.06 μm line from the YAG-laser might be used directly without frequency doubling to record on an extended red-sensitive Si-tube or a Newicon tube.

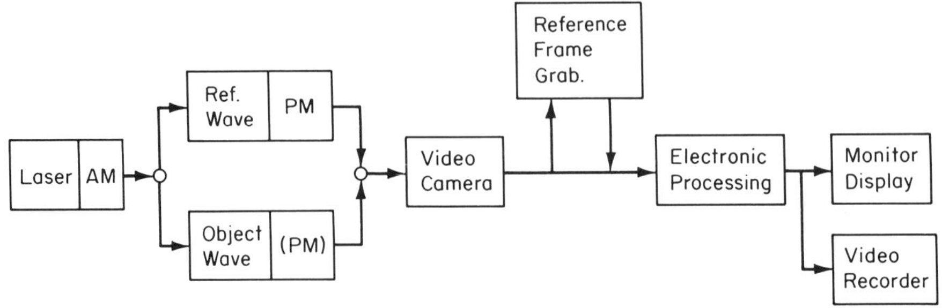

Figure 4.1. The elements of an ESPI-system.

More interesting might be that the wavelengths of most laser diodes coincide with the most sensitive part of the silicon cameras.

The AM-modulating block in Figure 4.1 means that the system can be run stroboscopically, double exposed within a video frame or used at shortened exposures. Again, laser diodes offer an interesting alternative as they are easily modulated.

Similar to other interferometers the light is split up into two branches, the object and the reference branch. In the object branch the illuminated test-object is imaged by a lens system onto the light-sensitive plate of the video camera or the target. The lens system is normally stopped down to about $f/30$. The reference branch provides a reference wave of uniform quality that impinges on the target from the same direction as the object wave. The two waves are directionally combined by an optical element that may be a simple wedge beam-splitter or more complex optical elements as described in Ref. 6. The reason for the stopped down lens and the in-line combination of the two waves is to keep the spatial frequencies in the primary hologram within the resolution of the video system. There is nothing magical, however, about this matching of frequencies to the video system's resolution and we might, for example, open the lens to increase the background light level of the interferogram. The added light, of course, acts as an incoherent addition. For a speckle reference system the aperture can be opened up considerably (8).

As shown in Figure 4.1, phase-modulation (PM) might be introduced into the reference wave (or into the object wave). This means that the phase of the wave is changed in a controlled manner either by reflection from a moving mirror or by transmission through a birefringent electrooptic crystal. The latter solution is more costly and demanding on the power supply, but has the advantage of reaching modulation frequencies in the MHz range. Facilities for phase-modulation are useful for vibration measurements and essential for fringe interpolation by phase stepping methods (9–11).

The video camera records the image hologram created by the combination of the object, and reference wave by transforming the spatial content of the primary hologram into an equivalent video signal. The video camera should have good sensitivity, large dynamical range, and high resolving power. These requirements are met to a very high degree by the modern multialkali tubes (like Chalnicon and Newicons) mounted in almost any type of respectable camera housing. However, conventional video cameras are—or will be replaced by charge coupled device cameras (CCD). For now the resolution of the commercially available CCDs are not quite good enough for the highest quality ESPI recordings for vibration testing. For deformation testing the resolution is determined by the digital store (512×512) and the better CCD cameras can be used with no loss of quality. The CCD camera is especially attractive for industrial applications due its robustness and immunity toward external magnetic fields.

The video signal from the camera is fed into an electronic processing unit that essentially performs a band-pass filtering and a rectification on the signal. This process can be shown to be entirely equivalent to the reconstruction process in ordinary hologram interferometry (12). For many applications we want to store a video frame to compare or subtract it from the frames coming from the video camera. Today, the video frame store is exclusively based on digital registration and storage. Note that the entire frame grabbing and the "reconstruction" process can (and will exclusively in the future) be performed digitally within the computer (13–14).

The modified signal is fed into the TV monitor where it is converted into a TV image, which can be considered as a reconstructed image hologram. As in ordinary hologram interferometry the image is interferometrically sensitive to any movement of the object and shows a fringe pattern indicating the amount of movement. The fringe patterns in ESPI have a coarser speckle structure due to the limited resolution of the video system compared with holographic film.

We like to point out some features that make the ESPI differ from ordinary hologram interferometry. The system produces 25/30 (European/American video standard) holograms a second at a normal exposure time of 40/33 mec. The high framing rate provides for great stability and tolerance to external disturbances. As the reference state can be updated at similar speeds we are able to experiment with objects undergoing rapid changes. In addition, the primary holograms are available as an electronic signal that can be relayed over great distances from the optical head (laser, video camera, and interferometer) to a surveillance center. The electronic state also allows for subtraction of holograms, which cleans up the image in terms of optical noise. Finally, digitalization of the electronic hologram opens up a great variety of signal manipulation, such as phase-stepping, and eventually provides full automatization of the system if desired.

Modes of operation

The ESPI system is used for the same measurement purposes as hologram interferometry, namely testing vibrations and deformations and for object contouring. The last application was strongly pursued by the Loughborough group, which provided intricate and new schemes for both absolute and relative contouring (6). (Note that the holographic illuminator used in their relative contouring system represents the basis for today's comparative hologram interferometry.) As the use of ESPI for contouring is of lesser interest at present, we do not delve deeper into this subject. We should, however, note that for a complete analysis of a complex vector field with subsequent strain calculation, information about the object topology relative to illumination and observation is essential (15).

Vibration testing

Due to simple instrument design, vibration analysis by ESPI is the oldest and most common application area. It is also most advanced in terms of image quality and acquisition of relevant information. The time-averaged mode of operation is most commonly used where the relation between the object's vibration amplitude $u_0(x_0, y_0)$ and the monitor intensity $I_{mt}(x_m, y_m)$ is given by a formula well known in hologram interferometry:

$$I_{mt}(x_m, y_m) = J_0^2[2\pi/\lambda \cdot g(\theta) \cdot u_0(x_0, y_0)] \tag{4.1}$$

where J_0 is a Bessel function of zero order and first kind while λ is the wavelength of the laser, and $g(\theta)$ is the usual geometric sensitivity factor associated with illumination, observation, and movement vectors.

An example of a time-averaged ESPI recording is shown on Figure 4.2. The brightest fringes represent zero order fringes, $J_0^2(0)$, and their centers are the nodal lines of the vibration pattern where the movement is zero. To convert the remaining fringes into isoamplitude contours we have to feed the fringe order into Equation 4.1 with the appropriate wavelength. If we assume $g(\theta) = 2$ (that is illumination and observation in line with the movement) and $\lambda = 0.633\,\mu$m (He–Ne laser), we find the first dark fringe to correspond to $0.12\,\mu$m, the first bright fringe to $0.19\,\mu$m, and so on.

If sinusoidal phase-modulation is introduced into the interferometric system, we observe the same fringe function. The fringe argument, however, depends on the vectorial difference between the object and the reference movement vectors, which makes the interferometer phase sensitive and provides an extended measuring range (16–18).

Let us take a look at what can be achieved from a properly constructed ESPI instrument with full facilities for vibration testing (obvious advantages common to hologram interferometry like no loading, no calibration,

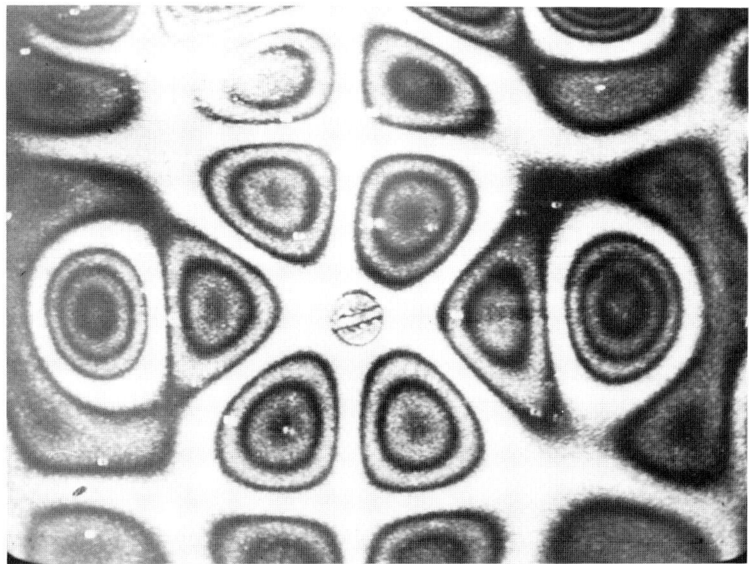

Figure 4.2. Vibrating steel plate recorded by time-average ESPI.

nondestructive testing, and so on are not listed):

1. The instrument will be compact and lightweight for easy transportation. Under normal building conditions there should be no need for working on a vibration isolated table.
2. Using the phase modulation with photoelectric detection the measuring range exceeds 10^6. A more realistic range by visual means would be from $5\,nm$ to $5\,\mu m$, where the upper bound is determined by the response of the phase-modulating element.
3. The phase-modulation principle used with a phase-shifting device provides vibration phase information as contours of constant phases, which is essential for analysis of complex movements (17) (or mode combination). Phase-modulation used with a slight frequency shift gives a dynamic presentation of the vibration pattern (18).
4. In principle there are no limits to the frequency range that can be inspected if the object can be sufficiently excited. The ability of the ESPI system to detect the very small amplitude encountered in the high frequency region means that higher frequencies can be reached.
5. The instrument will be tolerant against external instabilities. If a large laser, such as Ar laser, is used, the exposure time of each frame can be reduced, which gives an additional increase in stability (19).
6. For documentation purposes the appearance of the fringe patterns can be improved by speckle averaging (5, 20). If the averaging is performed digitally, the improved fringe pattern can be observed within a few

seconds. In the order of 100 fringes might be observable across the monitor image (note that this means absolute number of fringes, and not fringe order where the maximum number usually is more like 15 to 20).

7. The instrument can be used under high ambient light conditions, if a filter is incorporated we can look at objects in broad sunlight. The instrument may be used to study vibrating hot objects. Without any special arrangements we may measure the vibrations of objects heated up to the +1200°C range using a small HeNe-laser.

8. We have pointed out that the primary hologram is available as an electronic signal. This feature is important for digital treatment and computer interfacing. Many primary holograms can also be recorded by video tape or cassette recorder for later analysis. Finally, as the holograms can be relayed back by cable or wireless, there are hardly any limits as to how far away the optical head can be from the analyzing station.

9. The real-time operation of the system where 25/30 time-average holograms are presented each second makes it simple to analyze very complex vibrations within a reasonable time span. An example is the analysis of volume vibrations where simply rotating the object brings out the necessary information about the vibration vector (21).

10. Once the basic instrument is installed there are hardly any running costs apart from the electricity bill and documentation photography.

Thus far we have assumed the object to perform a sinusoidal vibration where the time average method combined with phase-modulation provides all the answer in an easily interpretable way. This method becomes complicated in use and eventually breaks down if the object vibrates at two or more frequencies simultaneously. Stroboscopic techniques have been used to analyze vibrations with two independent frequencies (21).

To analyze a general movement like a motor running under working conditions where several frequencies are present, we have to resort to pulsed lasers. Recent development of the repetitively pulsed YAG-lasers allow measurements under realistic conditions when used in an ESPI instrument. The analysis is complicated and this author feels that we still have a long way to go before meaningful and accurate answers can be presented within a short-time frame (in terms of computing time). This must be expected, however, as we are dealing with more complex movements.

Finally, we should note the possibility of combining the stopping action of YAG-laser with the derotator (22), which opens up for measurements on objects rotating at very high speeds possibly in the order of 100,000 rpms.

Deformation testing

Under this heading we include experimental conditions where we look at changes in the test-object that take place over a time period that spans two or more TV frames. Examples are objects being slowly deformed due to heating or undergoing a two-step change due to loading. This mode of operation makes it essential to record a reference frame by means of the memory store shown in Figure 4.1. This frame is to be subtracted subsequently from the frames coming from the video camera. The subtraction process provides the correlation fringes representing the deformation of the object between the subtracted frames. For video memories the older instruments relied on storage tubes (23), video tape recorders (24), and similar analog devices. Although giving fringe patterns of good quality these memory devices did not provide the easy handling and computer interfacing that are possible by today's digital frame stores.

The relation between the ESPI correlation fringes $I_{md}(x_m, y_m)$ obtained for a two-step change $u(x_0, y_0)$ is given by the same equation as in hologram interferometry:

$$I_{md}(x_m, y_m) = \cos^2[2\pi/\lambda \cdot g(\theta) \cdot u(x_0, y_0)] \tag{4.2}$$

(the terms have been defined earlier).

The deformation fringes represented by Equation 4.2 have constant visibility and we are able to observe higher fringe *orders* than for comparable time average recordings. As speckle averaging would demand a repetitive, identical deformation usually not encountered under practical experimental conditions, however, we have to contend with documentation pictures where the speckle carrier has full contrast.

It is reasonable to expect an ESPI system to resolve about 60 straight deformation fringes across the monitor image. In practical work such dense fringe concentrations are usually avoided by use of fringe compensation or simply updating again the reference state. Fringe compensation normally is achieved by moving or tilting a mirror in the illumination branch. Excessive compensation, however, may introduce fringe anomalies or contouring effects that may be mistaken for deformation fringes and should be used with some care. It is simpler and safer to update the reference state and follow the fringe build up anew, a technique that is essential for following rapid changes in the object. Note that we might also record primary hologram on video tape or cassette and perform the subtraction upon replay (25). In this way we can get deformation fringes between *any* hologram recordings if the magnitude of the deformation is within the resolution of the system.

Examples of ESPI deformation fringe patterns are shown on Figure 4.3. In Figure 4.3a a pressure transducer has been deformed due to differential pressure. Each fringe represents a deformation increment of $\lambda/2$ or 0.32 μm provided the conditions used in interpreting Figure 4.2 are also met. Figure 4.3b shows cracks due to increasing load in the center of a ceramic sample. The cracks are clearly seen as discontinuities in the fringe pattern.

If we try to list the advantages in using a commercial ESPI instrument for deformation testing like we did for vibration analysis, we find the items 1, 7, 8, 9, 10 listed previously to be directly comparable. In item 2 the measuring

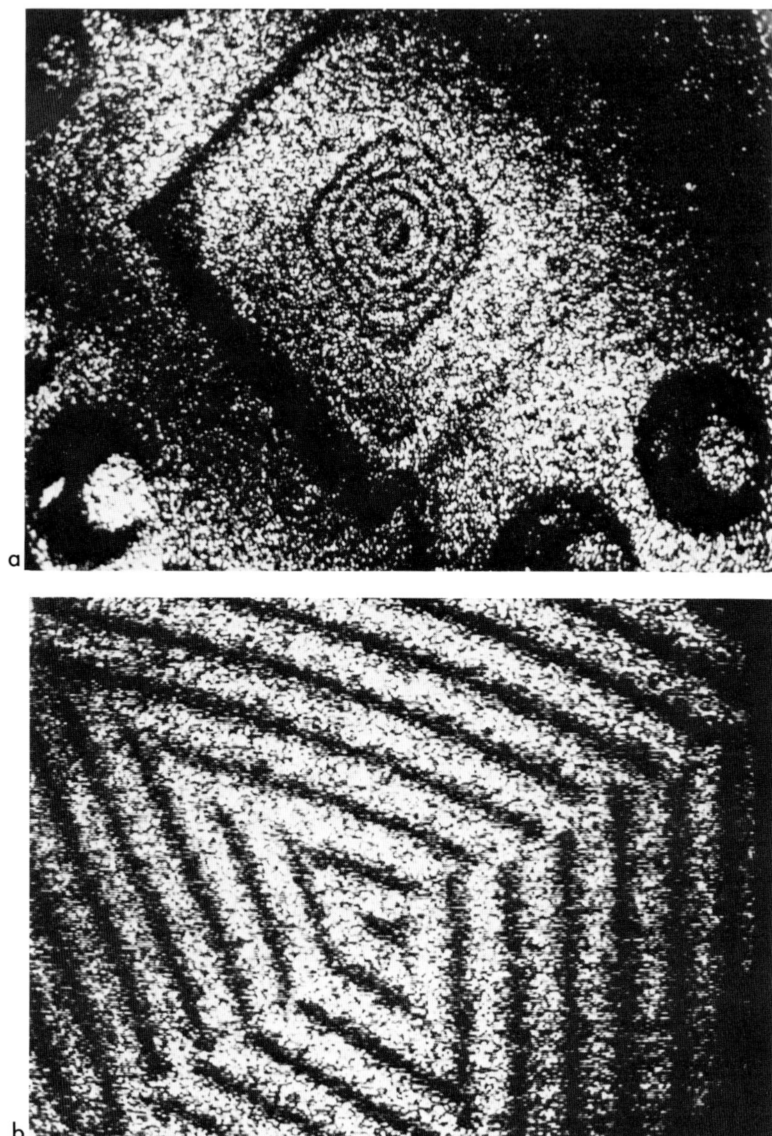

Figure 4.3. ESPI deformation patterns. a. Deformation of a pressure transducer. b. Crack pattern in ceramic sample.

range is smaller depending on the experimental conditions. A lower limit by visual read-out seems to be about $\lambda/25$ when straight reference fringes are added. The lower limit might possibly be improved by phase shifting and similar electronic fringe interpolation techniques (*note*: usually accompanied by a loss of resolution as we have to average over several speckles to improve the signal to noise ratio). Upward, the measuring range can be in the centimeter range if the deformation allows for continuously monitoring of the fringe build up. If the deformation is two-step in nature, we have to contend with the number of fringes indicated on Figure 4.3. In item 3, fundamental differences between ESPI-deformation testing from achieving the same stability as vibration testing (the system is still better than hologram interferometry). The only exception might be cases where the object to be tested is so big that the instrument actually can be clamped to the structure. In such experiments the stability is extremely good.

Miscellaneous

Here we discuss topics that are either common to both vibration and deformation experiments or deal with questions often asked regarding the technique.

Comparative ESPI

Due to the ESPI-subtraction process it is possible to compare directly the displacement of one object relative with another object. This may be potentially useful for rapid inspection of components, where the behavior of incoming components is compared directly with that of a standard component. This process can even be performed for vibrating objects, although the phase relation makes the results more difficult to interpret (26).

Shear ESPI may be considered to be a special kind of comparative ESPI where we compare the test object with itself in a spatially shifted version (27–29). As in conventional interferometry the shear is usually in the x and y directions along the object's surface, although, for example, radial shear might be useful for certain purposes. Using (a small) shear in the x and y directions gives the second derivative of the out of plane displacement, which reduces the mathematical derivation of bending strain by one order. As shearing interferometer can be considered common path interferometers, they are very stable and should be considered for work on unstable objects. The main drawbacks with shearing interferometry is that the interpretation of the fringe pattern may not always be obvious and in addition the sensitivity is low when a small shear is used. However, computer interpretation and phase shifting techniques may be a big help here.

Object size

The size of an object that can be inspected depends on two different aspects, how much light is reflected back to the instrument from the object and how small details we want to observe.

The first requirement is dependent on available laser power and object reflectance. If a 5 mW laser is used, we normally would say that a 30×30 cm^2 white, matte surface would represents a maximum area for recording vibration fringes (deformation recordings are less critical due the constant intensity of the higher order fringes). If retroreflective paint is used on the object, the observed area can be in the order of 1 m^2, whereas retroreflective tape increases that number to about 7 m^2.

The resolution of the system is most critical for deformation analysis because speckle averaging techniques cannot be used and the digital resolution is at present only 512×512 pixels. In addition the dynamic display provided by phase modulation increases the spatial detection limit. If for deformation testing we assume 20×20 pixels to be a reasonable number for positive visual detection of a fringe anomaly, we find that for a 1-m^2 object defects down to about 4 cm (linear size) can be found. If we want to detect 0.5 cm defects, we need to reduce the observed area to about 12×12 cm^2 and so forth. These numbers vary with the actual set-up, the object, and the type of defect and may also be improved by computer-assisted read-out. However, ESPI cannot and will not, even in the future, compete with ordinary hologram interferometry in terms of large coverage combined with high resolution. This is especially true for deformation testing. It is the speed and flexibility of the ESPI system that makes it interesting for practical work.

Automated read-out

Computer-assisted read-out and automatic fringe analysis are essential for industrial purposes and there is a great activity in this field in addition to the recent reports (9–11). It is also of interest to be able to improve the fringe accuracy for strain calculations. At present we do not know what accuracies we might expect, maybe about $\lambda/100$ for deformation measurements and $\lambda/500$ for vibration measurements by various phase shifting techniques. (Lower values for vibration measurements have already been reached (18), however, these values have been reached by spot measurements on the monitor and lock in averaging.) These numbers depend greatly on the experimental conditions.

We should note that by phase shifting techniques, several frames have to be read sequentially into the computer. In addition, we might loose a couple of frames between each phase step to allow for stabilization of the phase stepping device. This effectively leads to a stricter demand on the stability of the system, the object, and the surroundings and we loose one of the stronger advantages of the TV–holography system.

In general, the computer can easily be incorporated into the ESPI system, for example, to write experimental data on the video monitor for documentation purposes or to monitor experimental parameters, such as frequency, phase, and excitation level.

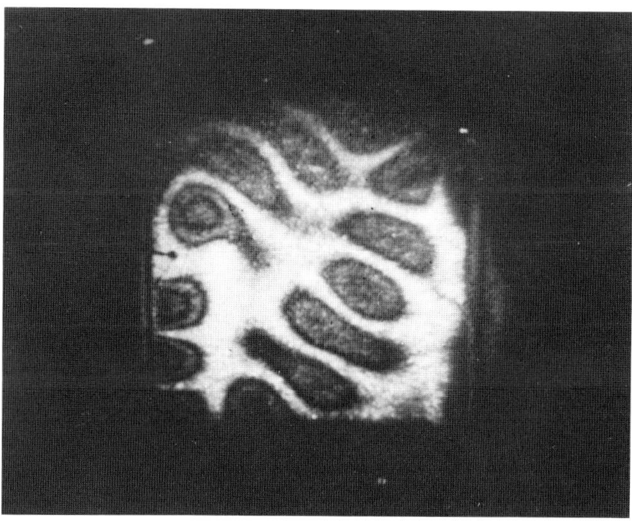

Figure 4.4. Vibrating object placed outdoor 65 m from ESPI instrument.

Remote inspection

We have already suggested that the optical head can be placed at large distances from the monitoring station due to the electronic representation of the primary holograms. It is also possible to use the instrument to inspect objects that are placed at considerable distances from the optical head (30) as shown in Figure 4.4. To inspect difficult, accessible places, such as cavities, fiberoptic imaging and illumination works very well in conjunction with ESPI (31). An interesting extension is provided by using monomodus fibers to lead the light from a strong central laser to the optical head that is moved around in a factory environment (32).

Applications

As already stressed, we perform the same measurements by ESPI as we do by hologram interferometry. The speed and stability of the system, however, allow us to tackle experimental difficulties that would either have been extremely difficult by hologram interferometry or would have called for a pulsed laser.

So far vibration analysis has been the most common application of the technique and where most impressive results have been achieved both for medical and industrial purposes. Within the medical field, the vibrations of the ear drum of humans have been investigated in vitro and in vivo (33). The basilar membrane (in vitro) represents another challenge that has been attacked with a high degree of success (34). If global interferometric

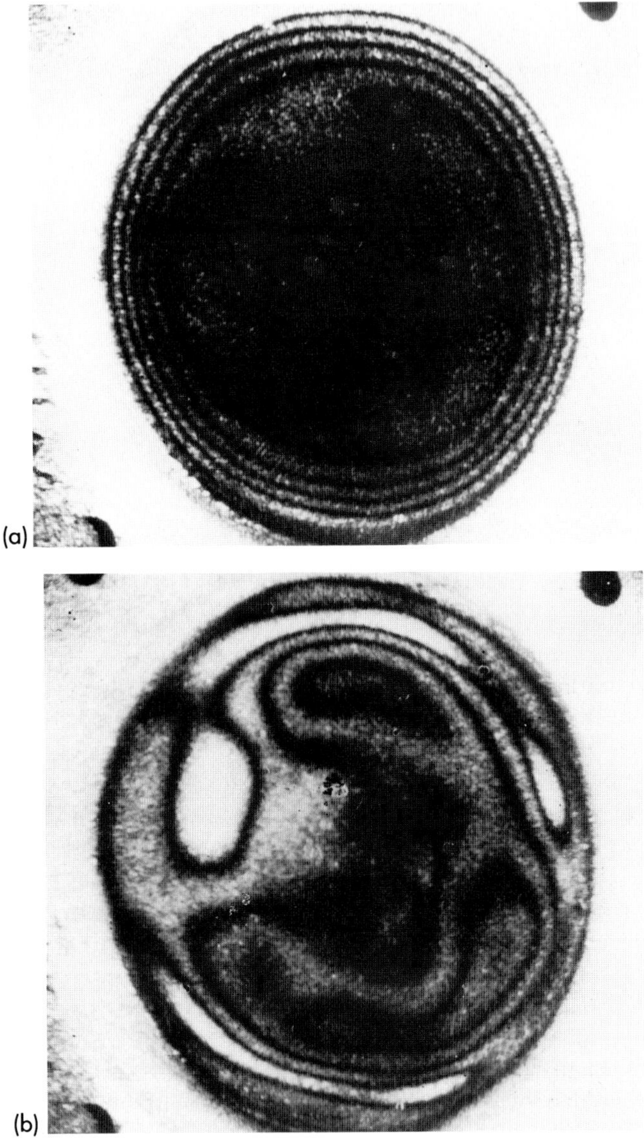

Figure 4.5. Vibration patterns of a loudspeaker at (a) low frequency, (b) medium frequency, and (c) high frequency.

methods ever progress from the experimental medical stage to the practical clinical usages, then ESPI will play an important role.

The testing of turbine parts represents a typical example of an ESPI-application area in the industry. An ESPI instrument constructed for a Norwegian gas turbine manufacturer 10 years ago was used on-site to check the resonances of each blade (35). The rather primitive set-up was used by

(c)

Figure 4.5 (*continued*)

ordinary mechanics during production and the system was claimed to have repaid its cost within the first 2 months. The new compact commercial version of the ESPI instrument, RETRA 1000, has been in use at the same plant over a year without problems.

Vibration analysis of loudspeakers represents another industrial application area and some examples of vibration on patterns are shown on Figure 4.5a, b, and c. The phase distribution is essential for understanding the complex movements often found in loudspeakers and this information is acquired far easier by ESPI than by other methods. In the example shown previously, the recording at medium frequency would need phase information for interpretation as the appearance of nodal points is a sure sign of mode combination that sets up a traveling wave across the surface.

The vibrations in various components of a car represent an important field of interest. Figure 4.6a shows a car door (Mazda 818) excited by sound pressure, as recorded by the RETRA 1000 instrument with the built in 5-mW laser. The surface was painted with retroreflective paint to improve the reflectance. Note the peculiar vibration pattern on the lower part of the door where the nodal areas have degenerated to nodal point. Again this pattern is typical of combination modes where waves travel across the object. This complex movement could be broken down into contours of constant phase and amplitude using the phase sensitivity of ESPI. In Figure 4.6b we are observing the car door from the same position, but we have changed the magnification to look at the area around the door lock, which at this particular frequency had a local resonance.

In Figure 4.7 we are using the same instrumentation to look at an entire

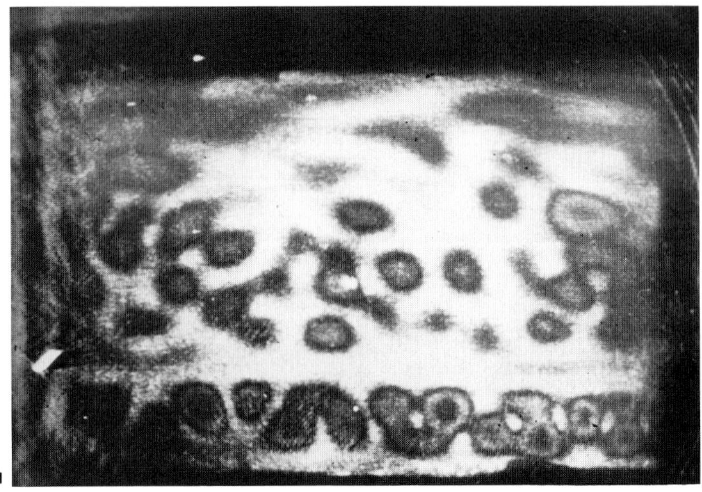

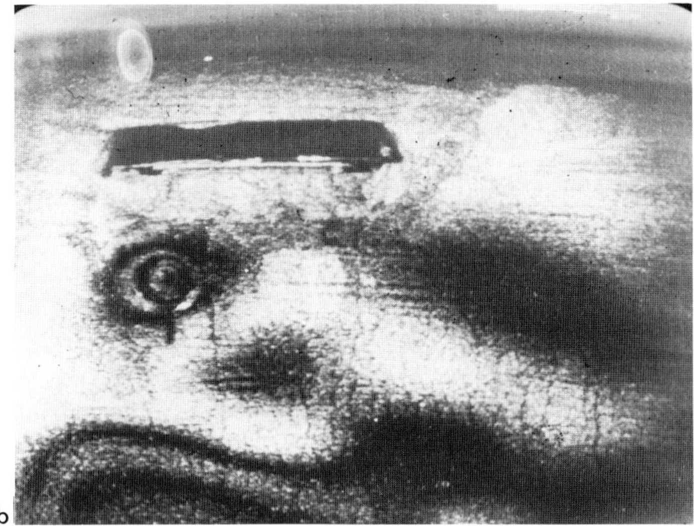

Figure 4.6. Vibrations of car door. a. The entire area. b. Increased magnification observing the lock area.

car of a well-known brand. The car body is excited by a loudspeaker placed in the front seat. In this case the body was covered with a sheet of retroreflective tape that gives a further increase in the reflectance. The tape is very thin and lightweight and does not change the resonance vibration of this particular object. The investigations were performed both indoor and outdoor under rather harsh conditions (36).

Deformation testing by ESPI seems to have been mainly confined to industrial research, where one area of interest is the detection and

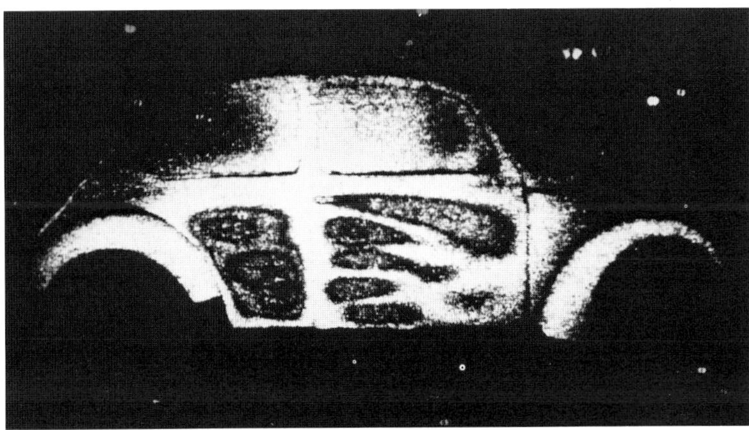

Figure 4.7. ESPI recording of an entire car body (courtesy of Malmo and Vikhagen).

monitoring of cracks in various materials as the loading increases. Experiments have been performed on concrete samples where dramatic changes in the material behavior have been observed as the loading increased. Information found in this way has proved important for determining the long time strength of concrete.

This type of deformation testing is also of particular industrial interest for ceramic materials. An example on how we can detect and measure crack length and propagation was given in Figure 4.3b. Ceramic materials are used at high temperatures and we have done ESPI measurements at temperatures well above 2000°C. In fact, the technique has proved able to record the behavior of small objects heated to temperatures over 3000°C using a medium power Ar laser. Working on steel samples we have been able to detect phase changes and measure the growth of oxidation crusts (37).

We have also used ESPI to inspect composite materials for internal defects and the technique appears to be well suited for detecting most interior defects (38,39).

Figure 4.8 shows a honeycomb panel (actually a region on a microwave antenna) with a small interior defect caused by crushing of some cells. The sample was heated for a short time on the back side while observed from the front side. Figure 4.8a shows the fringe pattern recorded right after the hot air current has been turned off. The surface bulging caused by the faster heat transfer through the defect is shown as a local fringe center. The coarse, slightly curved fringe pattern is caused by uneven heating and a tilting of the antenna. Figure 4.8b shows the fringe pattern a few seconds later when the defect has been largely masked by the bulging over the individual cells. This is a rather standard experiment that could also have been performed by ordinary hologram interferometry with some care. Here,

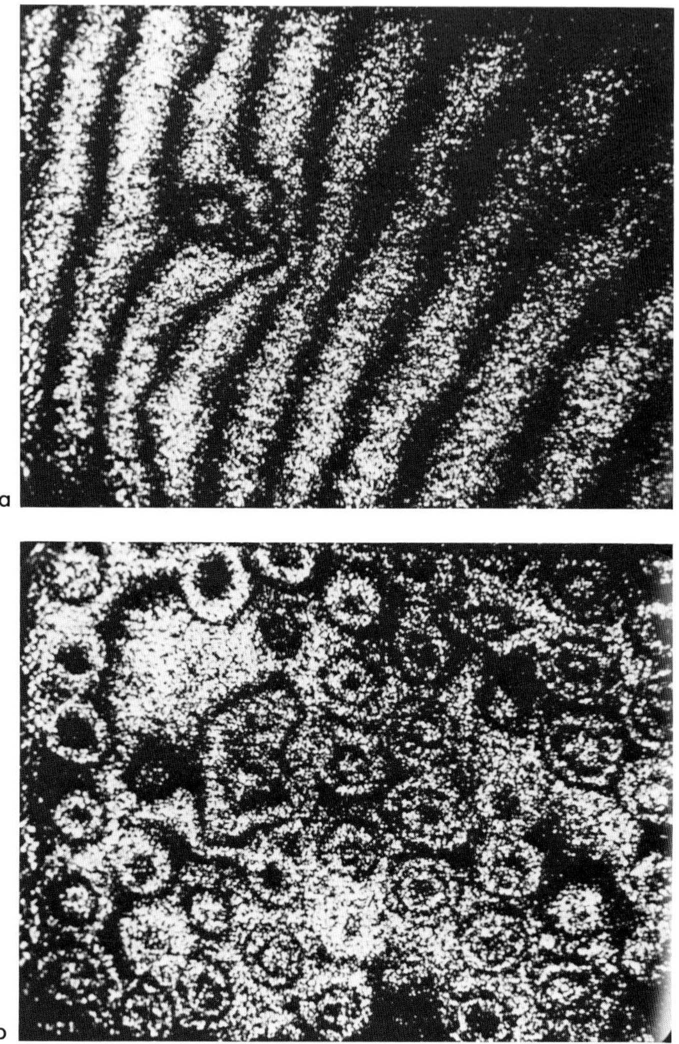

Figure 4.8. Defect detection in a honeycomb panel by heat propagation. a. Immediately after heating. b. Some seconds after the heating.

the detection is really based on the heat being conducted faster through the defect region than the good cells. Therefore, a dynamic technique, like ESPI, where the reference state can be easily updated has a definite advantage. In fact, watching a live progression of the experiment sampled in Figure 4.6 gives a much more convincing and positive identification of the defect. Certain defects in laminated structures have been found only after using the technique's ability to show transient behaviour.

Concluding remarks

We have described the present state of ESPI and given some examples of its applications. It is tempting to dare some predictions about developments we can expect regarding the instrumental part of ESPI and the technique's field usage within the next decade.

The future laser source in the instrument may be run both continuously or repetitively double-pulsed synchronized to the TV-framing rate. The laser will probably be a laser diode or a diode pumped YAG, both sources being compact with high power and coherence. The recording media will be CCD or similar devices, where a resolution of at least 1024×1024 pixels will be standard. The computer will be an integral part of the instrument providing speckle-free fringe patterns and relevant numbers for vibration and deformation parameters in three dimensions. The measurement accuracy has improved, but is not good enough to compete with electronic strain gages? For on-line defect detection and similar inspection, fully automated instruments will be available. The instrument itself will be compact and battery-powered for full freedom of transportation and use, indoor as well as outdoor. A complete instrument (laser, integrated optics, camera, computer, flat screen monitor, and a compact disc (?) for data storage and read-out) might have a total size that compares to the movie cameras of today.

The future application areas are hard to predict. The technique should eventually find its way to major research laboratories. Here, it will be used within developmental work and testing of components of all sizes and under all conditions.

As for industrial inspection in the field, definitely it will be possible to detect weak spots in structures and find their level of strain. These measurements might be performed from a distance, or riding the instrument on the structure or in cavities by fiber optics. There are numerous applications that seem to be possible, but a massive effort is necessary to gain experience and interpret the results. This is especially true for work within the biomedical field, where application areas have hardly been scratched and where a dynamic system like ESPI is bound to give exciting results.

The question remains if the competition from fast, reusable holographic medium, like the BSO-crystal (40, 41), will slow or stop the developments indicated previously. Hologram interferometry definitely has, and will continue to have, the edge when it comes to resolution and image quality. It is hard to envision that the TV system as we know it today may come anywhere near the information content of a holographic plate. When it comes to surveillance of rapidly changing events or on-line inspection, we end up looking at the fringes with a video camera. In such cases we might as well use the video camera to record the holograms directly, especially, if the analysis is to be carried out in a computer with a limited number of resolution cells the same.

Acknowledgment

The work described in this chapter has been partly supported by the Norwegian Council for Industrial Research (NTNF). The author also wants to thank the CONSPECTUM company for kindly providing the RETRA-1000 instrument for most of the experiments.

References

1. Macovski, A., Ramsey, D., and Schaefer, L. P. *Appl. Opt.* 2722 (1971).
2. Butters, J. N., and Leendertz, J. A. *J. Meas. Control.* **4,** 349 (1971).
3. Schwomma, O. *Oesterreichisches pat. no. 298830* (1972); see also Neuer Zuricher Beilage "Forschung und Technik" (in German) 257 (1975).
4. Kopf, U. In *Messtechnik.* (in German) **4,** 105 (1972).
5. Slettemoen, G. A. *Appl. Opt.* **19,** 616 (1980).
6. Jones, R., and Wykes, C. *Holographic and Speckle Interferometry.* Cambridge U.P., London (1983).
7. Løkberg, O. J., and Slettemoen, G. A. *Applied Optics and Optical Engineering.* (R. R. Shannon and J. C. Wyant, eds), Vol. X, pp. 455–504 (1987).
8. Wykes, C., Butters, J. N., and Jones, R. *Appl. Opt.* **20,** A50 (1980).
9. Nakadate, S., Yatagai, T., and Saito, H. *Appl. Opt.* **24,** 2172 (1985).
10. Creath, K. *Appl. Opt.* **24,** 3053 (1985).
11. Robinson, D. W., and Williams, D. C. *Opt. Comm.* **57,** 26 (1986).
12. Løkberg, O. J. *Phys. Technol.* **11,** 16 (1980).
13. Nakadate, S., Yatagai, T., and Saito, H. *Appl. Opt.* **19,** 1879 (1980).
14. Creath, K. *Proc. SPIE.* **501,** 292 (1984).
15. Winther, S. *Opt. and Lasers in Engr.,* **8,** 45 (1988).
16. Løkberg, O. J., and Høgmoen, K., *J. of Phys. E.,* **9,** 847 (1976).
17. Løkberg, O. J., and Høgmoen, K. *Appl. Opt.* **15,** 2701 (1976).
18. Høgmoen, K., and Løkberg, O. J. *Appl. Opt.* **16,** 1869 (1977).
19. Løkberg, O. J. *Appl. Opt.* **18,** 2377 (1979).
20. Løkberg, O. J., and Slettemoen, G. A., *Proc. ICO-13,* p. 116 (1984).
21. Løkberg, O. J. *Opt. Eng.* **24,** 356 (1985).
22. Brevig, B. Vibration Measurement by ESPI and Acoustic Frequency Analysis. Ms. thesis. 1985 (In Norwegian).
23. Fagan, W. F. *Proc. SPIE* **398,** 193 (1983).
24. Løkberg, O. J., Holje, O. M., and Pedersen, H. M. *Opt. Laser Technol.* **8,** 17 (1976).
25. Butters, J. N., In *The Engineering Uses of Coherent Optics* (E. R. Robertson, ed.) Cambridge University Press, Cambridge, pp. 155–170 (1976).
26. Winther, S., Løkberg, O. J., Slettemoen, G. A., and Malmo, J. T. submitted to *Opt. Engr.*

27. Løkberg, O. J., and Slettemoen, G. A. *Appl. Opt.* **20,** 2630 (1981).
28. Leendertz, J. A., and Butters, J. N. *J. Phys. E: Sci. Instr.,* **6,** 1107 (1973).
29. Nakadate, S., Yatagai, T., and Saito, H. *Appl. Opt.* **19,** 424 (1980).
30. Slettemoen, G. A., and Løkberg, O. J. *Appl. Opt.* **20,** 3467 (1981).
31. Løkberg, O. J., and Malmo, J. T. *Opt. Engr.* **27,** 150 (1988).
32. Løkberg, O. J., and Krakhella, K. *Opt. Comm.* **38,** 155 (1981).
33. Davies, J. C., Buckberry, C. H., and Jones, J. C. D. SPIE-proceeding #863 "Industrial Optoelectronic Measurement Systems Using Coherent Light" Cannes, 16–20 Nov. 1987.
34. Løkberg, O. J., Høgmoen, K., and Holje, O. M., *Appl. Opt.* **18,** 763 (1979).
35. Neisswander, P., and Slettemoen, G. A. *Appl. Opt.* **20,** 4271 (1981).
36. Løkberg, O. J., and Svenke, P. *Opt. and Laser Eng.* **2,** 1 (1981).
37. Malmo, J. T., and Vikhagen, E. *Exp. Techn.* **4,** 175 (1988).
38. Løkberg, O. J., Malmo, J. T., and Slettemoen, G. A. *Appl. Opt.* **24,** 3167 (1985).
39. Løkberg, O. Int. Conf: Post Failure Anal, Techn. for Fiber Reinforced Composites. pp. 13/1–13/9, Dayton, Ohio, (1985).
40. Løkberg, O. J., Malmo, J. T., and Strand, A. AGARD Proc, No. 402. The Repair of Aircraft Structures Involving Composite Materials, pp. 7/1–/7/7. Oslo, Norway (1986).
41. Huignard, J. R., and Herriau, J. P. *SPIE* **215,** 178 (1989).

II
Nondestructive testing

5
Holographic, nondestructive evaluation of photovoltaic cells

LARRYL MATTHEWS, GEORGE MULHOLLAND,
and GINA SADA RIGHTLEY

Evaluation of the mechanically and thermally induced stresses in a photovoltaic cell by holographic interferometry is not a new idea. Past investigations, however, relied heavily on applying stress to the individual cell by mechanical means and the results met poor acceptance. In addition to some of these relatively unsuccessful early holographic tests on photovoltaic cells, investigators have been skeptical of using holography to study internal defects, because only surface displacements are actually monitored. Current favorite techniques for studying internal defects are x-rays and ultrasonics. Although these methods are important tools and offer ample cell structure information, holographic nondestructive evaluation (HNDE) has the potential to add to the viably obtainable cell and assembly stress characteristics.

In this report the successful use of holographic interferometry, from now on referred to as holometry, in studying potential or actual photovoltaic cell mechanical failures is described. Holometry has shown itself to be a useful tool for evaluating photovoltaic cell behavior. Indeed, holometry does have its drawbacks, but the added features brought to the study of photovoltaic (PV) cells outweigh the drawbacks.

A schematic of the cell and the cell assembly design used in this report may be seen in Figure 5.1. The substrate on which the whole assembly is built is composed of a thin, 2-inch square wafer of alumina. This wafer has been metalized with copper on the areas to which the cell and the lower buss bars are attached. The actual PV cell is a silicon cell approximately 1-inch square that is soldered on to the copper metalization. Two electrodes are seen in Figure 5.1. In operation, the upper electrode pulls current off at the top surface of the cell by use of the interconnects, which are soldered to the cell itself, and the upper buss bar. The lower electrode pulls current off of the lower surface of the cell making use of the metalized (copper) coating, between the alumina and the cell proper, and the lower buss bar. The buss bar assemblies are composed of tinned copper. This information is supplied to aid in an understanding of the cell and the cell assembly. The

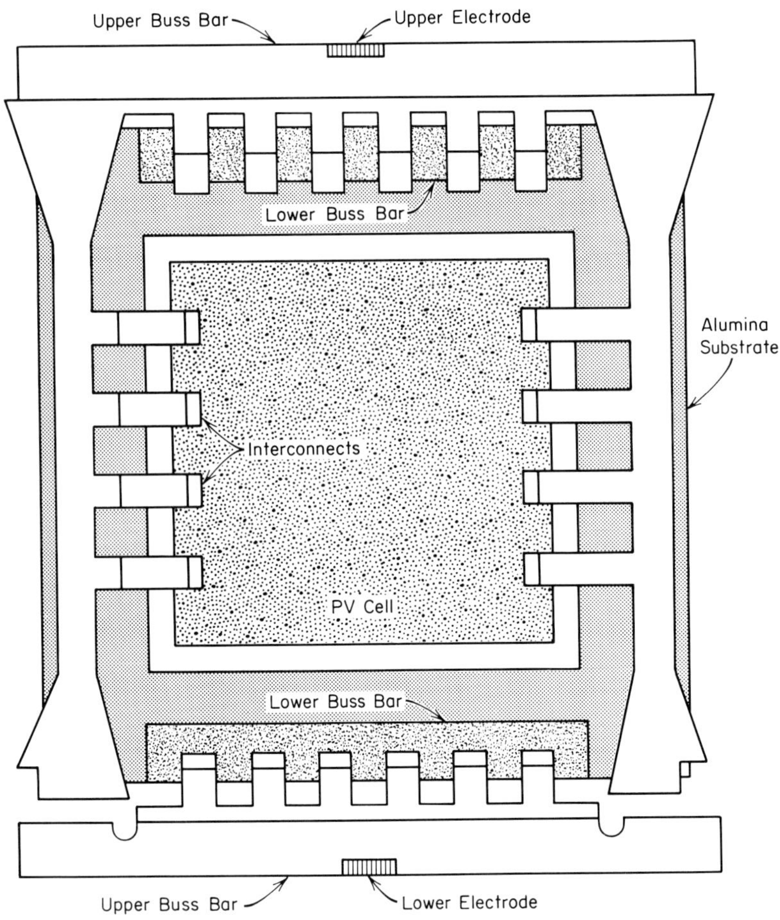

Figure 5.1. A schematic of the cell and the cell assembly design.

actual operation of these cells in not the concern in this report, but rather the determination on detecting of potential failure areas and stress complications are discussed.

X-ray analysis of PV cells (Fig. 5.2) can show areas of differing density within a cell layer and can do so with a noticeable amount of detail.

X-rays are relatively easy to make once the appropriate equipment has been purchased and successful shielding of the human participants has been achieved. The use of such a system tends to be expensive, and the results do not supply information about the interaction between the PV cell, the interconnects, and the alumina substrate. In addition, x-rays cannot study or identify problem areas in cell assembly designs that are revealed only under stressed conditions.

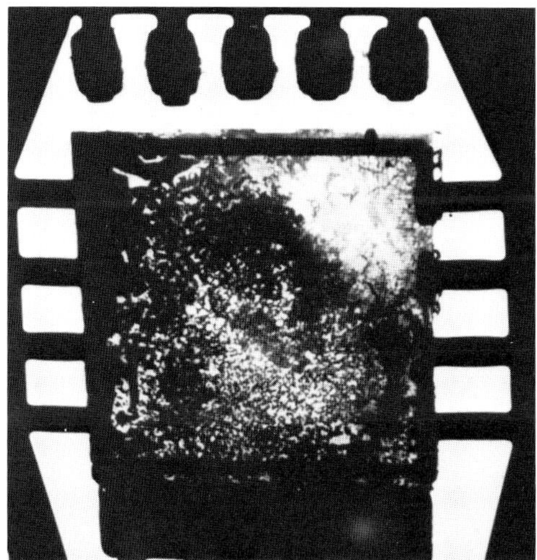

Figure 5.2. X-ray of a PV cell. The upper right corner shows a region of different density by virtue of its light color (courtesy of Dr. Robert Barlow, Sandia National Laboratories, Division 6221).

Ultrasonic analysis is also capable of supplying the same kind of information available from x-ray techniques. The resolution is slightly different and takes a slightly different format, but is nonetheless useful for studying PV cell defects. Figure 5.3 is a photocopy of a color photo supplied by Dr. Robert Barlow from an ultrasonic test of the same PV cell discussed earlier and shown in Figure 5.2. Unfortunately, much of the detail is lost in the black and white reproduction but the relative high quality information is still evident. The upper right hand corner shows an even greater deviation from the rest of the cell than shown in Figure 5.1. In addition, there is a little more detail around the cell interconnects than available from x-rays.

Ultrasonic analysis is clearly capable of showing some of the potential problem areas within a PV cell. An ultrasonic system, however, is limited to studying PV cells at or near room temperature, although some slightly elevated temperature analysis is possible if enough effort is expanded. In addition to the problem of not being able to study PV cells at elevated temperatures or during thermal cycling, ultrasonic analysis is not capable of showing the intricate detail of the interactions between cell interconnects and the cell surface.

As written in the statement of work for the Sandia contract, which provided funding for this study,

> The purpose of this effort is to develop holographic interferometry as a nondestructive testing method for photovoltaic concentrator cell assemblies and to provide information that will allow the assessment of the potential of holographic interferometry as a quantitative diagnostic technique.

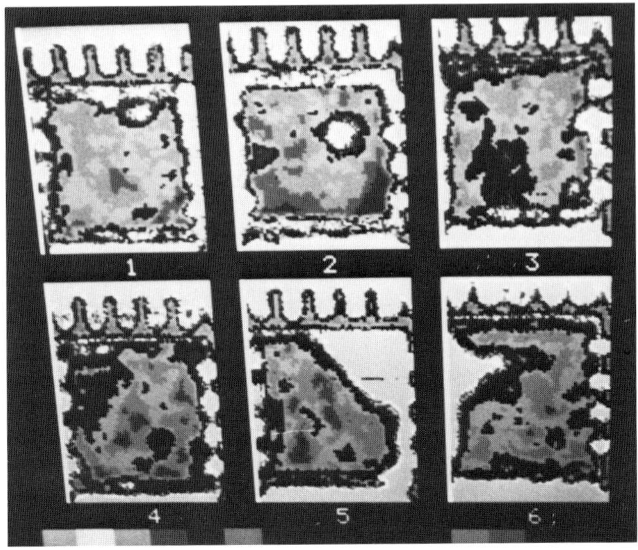

Figure 5.3. Ultrasonic photo (courtesy of Dr. Robert Barlow).

The remainder of this report discusses the steps taken to evaluate PV cell mechanical performances under varying conditions using holometry.

Testing procedures

Several mounting and heating testing procedures were tried to determine which would provide the highest quality information and the most repeatable information. In addition to the PV cell mounting concerns, the holographic system was rearranged and modified to provide the greatest sensitivity for studying a given part of the cell. With the holographic system, a lens can be used to magnify various parts of the PV cell for closer study, which, in some cases involved multiple lenses to reduce monochromatic aberrations while increasing magnification.

Figures 5.4, 5.5, and 5.6 show the PV cells mounted in an oven and free-standing on the optical table. Tests were conducted using both configurations with the oven supplying a uniform heating environment for the PV cell, and a spot heater was used to supply thermal heating to the front of the cell only. The oven was heated from ambient to approximately 150°C, and the spot heater supplied maximum heat flux of $20\,\text{W/cm}^2$. The temperature of the cell depended on how close the spot heater was positioned to the cell and how long the heat was applied. Subsequent holometric testing of the PV cell showed that visualizing defects and high deformation areas did not change appreciably with time or spot heater position, therefore, the maximum flux was not needed to test the cell.

In addition to the two methods used to heat the PV cell, two mounting methods were tried both in and out of the oven. One method required using the mounting structure supplied with the cells. This structure is an aluminum plate with cooling fins on the back and two clamps attached to the front. These clamps held the PV cell down on the plate by attaching to the alumina substrate on the back of the cell. The cell and the aluminum-finned structure can be seen in Figures 5.4, 5.5, and 5.6. Besides the aluminum-finned structure, a special invar structure was designed and fabricated by Dr. Bruce Hansche at Sandia National Laboratories to hold the PV cells in such a way as to reduce greatly the effect of the mounting, therefore, allowing the cell to be studied irrespective of the way it will be mounted eventually.

Besides the mounting and heating methods that were studied in this investigation, several optical configurations were tried before choosing the arrangement shown in Figure 5.7. The holographic medium is a reusable thermoplastic plate marketed by Newport Research Corporation. The medium can be erased, exposed, and developed in situ within approximately 2 minutes. This allowed several illumination and exposure sequences to be performed over a short time span. In addition, the PV cells could be changed and remounted rather easily to allow several techniques to be tried. Over the course of the study, several thousand holograms were made using the thermoplastic plates. Each plate produces high quality holograms over approximately 500 exposures.

Figure 5.7 also shows that the reference beam was launched down an optical fiber. It was necessary to use an optical fiber while using the oven as the distance between the oven and the holographic camera did not allow enough room to redirect and then expand and collimate the reference beam using lenses. Once the fiber was used, it became so convenient that it was used for every other configuration as well. This fiber is made of a high birefringent, polarization preserving material that was cut and terminated in the Optics Lab at NMSU. The fiber illumination technique worked very well and continues to be used in the laboratory.

In addition to the holographic camera and the optical fiber illumination technique, notice the lens shown in Figure 5.7. As mentioned earlier, different parts of the PV cell could be magnified and studied using the lens. Many different combinations of single lens magnifications were tried to determine how best to view the PV cell. For single lenses, even though the single lenses are actually cemented doublet achromats, trade offs had to be made between magnification and monochromatic aberrations. As the cell area was magnified greater and greater, the distortions of the field of view also became greater. The lenses chosen as the best compromise to perform

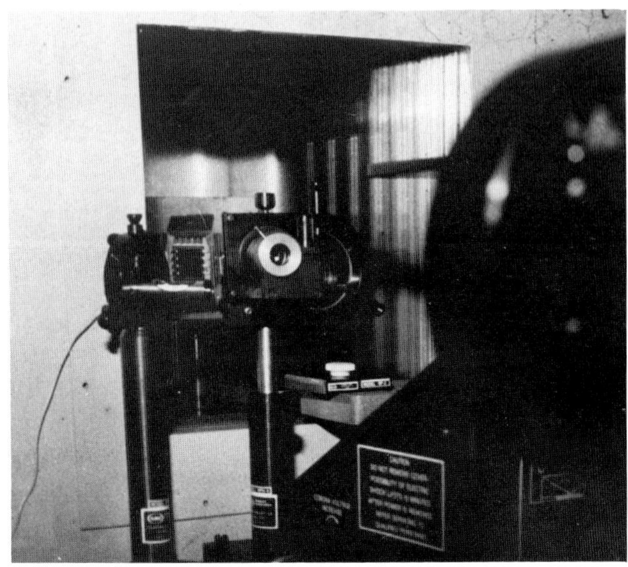

Figure 5.4. Oven testing configuration.

the task at hand had focal lengths in the 100- to 200-cm range. Toward the end of the study, greater magnification was desired and multiple lens combinations were tried.

Returning to the mounting and heating procedures, after several combinations were tried, the nonuniform heating and the aluminum plate mounting structure were chosen for the detailed study of the PV cells. The

Figure 5.5. Oven testing configuration—alternative view.

Figure 5.6. Free-standing testing configuration.

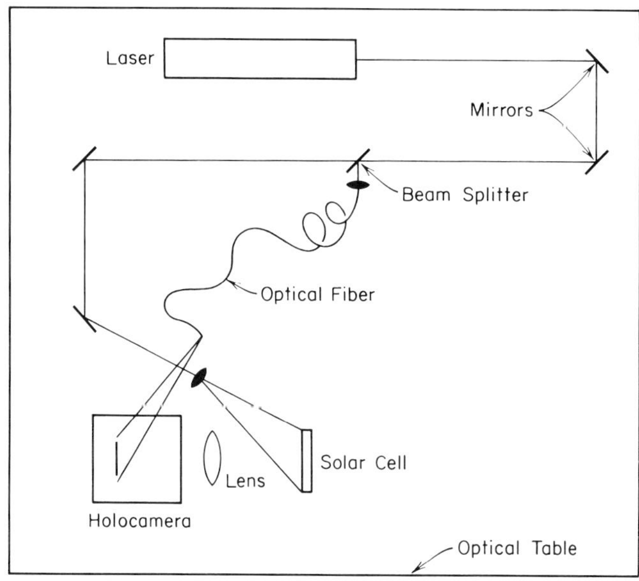

Figure 5.7. Optical bench set-up.

basic reasoning for these selections proceeds as follows. The nonuniform heating (using a spot heater) procedure was chosen because it is closer to the thermal environment that the actual PV cell will see when in use. Also, the holometric results obtained using this procedure appeared to be more repeatable, reliable, and easier to obtain than using the oven configuration. Certainly the spot heater allows for much quicker turn around on testing the PV cells, and it is easier to use. This could be especially important if such a procedure was to be used in a production environment or on testing of several PV cells over a short time frame.

The constrained aluminum plate mounting structure was chosen over the special invar structure because it provides the actual mount situation experienced by the cell assemblies in the field, and it was found that the mount with the constrained aluminum plate, provides more overall information about the cell. It can certainly be argued that the unconstraained configuration could be used to study the PV cell in a more pure form, free of external influences, however, several hundred holograms of experience has convinced us that the cell defects and other deformations not related to defects are more dramatic and more easily seen by using the constrained mounting. This strongly implies that any mounting structure used with the PV cells in the field should be studied in the laboratory to determine what effect it has on the mechanical structure of the cell.

Flaw detection

As mentioned earlier, holometry can be used to study many different aspects of the mechanical interactions between various parts of a PV cell. Before getting into the details of studying the interconnects, consider a comparison of the same PV cell. Figure 5.8 shows that PV cell. Unfortunately, these photos were taken early in the study while the photographic techniques needed for recording the hologram on a 35-mm format were still being developed. However, the delamination at the upper right corner is still visible. The obvious bumps in the otherwise smooth bull's eye patterns indicate that a flaw is present. Because deformations in the cell are orthogonal to the fringe pattern, the delamination motion appears to be centered about a point midway between the right and left side of the PV cell and close to the top of the cell just slightly above the center of the bull's eye. Of course, the actual hologram was much sharper than the photographs.

While viewing the figures, the fringe pattern can be thought of as a contour map with the distances between fringes representing fractions of a micrometer. The delamination appears as a ridge moving down the mountain, although the pattern could just as easily represent a valley as the difference is not easily discernable from the hologram. Besides the delamination, more useful information can be obtained from these figures.

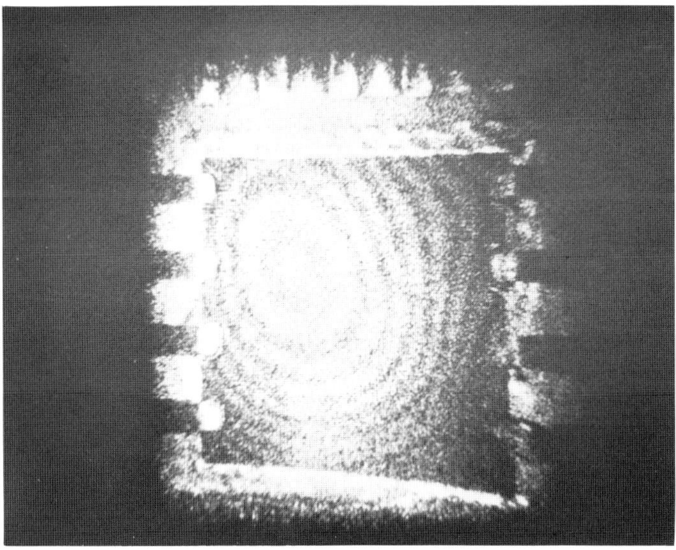

Figure 5.8. Delamination hologram.

First, in each case the center of the bull's eye, representing the top of the hill or the bottom of the valley, is located in the upper left quarter of the PV cell active area. This is true in spite of the source heating being applied in the center of the cell and in fact sometimes on the right side of the cell. Each PV cell tested had a slightly different position for the center of the bull's eye. This tends to indicate that the cooling of the cell surface by the substrate is slightly different for each cell. For all of the cell assemblies tested, it was determined that the bull's eye represented a valley rather than a hill. This information was obtained by use of a real-time hologram. The previously stressed cell was touched with a pin in the center of the bull's eye. This caused the number of fringes to increase indicating that the direction of movement caused by the pin was the same as that caused by the thermal stress. In addition to cooling mechanisms, the way the cell is mounted or rather the mounting structure itself plays a significant role on how the displacement patterns are established by the cells.

One other potential flaw in the cell can be detected by a closer study of Figure 5.8. When these photos were taken the importance of looking at the interconnects was not completely appreciated and, therefore, no concentrated effort was expended on the study of this cell's interconnects. After performing several such tests toward the end of the study, however, experience showed that potential problems can be seen from old pictures. In Figure 5.8 the second interconnect from the top on the left side appears to be affecting the fringe pattern on the PV cell active surface. This is a subtle effect in this photo, however, a much more dramatic example will be shown later. Unfortunately, the cell used to generate the fringes shown in Figure 5.8 was returned to Sandia before a close-up study of the interconnect region could be performed.

A comparison of Figure 5.8 with Figures 5.2 and 5.3 shows that some detail concerning the internal constitution of the photovoltaic cell is not as evident from the hologram, as it was from the x-ray and ultrasonic techniques. Displacements and interconnect information, however, obtained during heating from the hologram is not available from the other techniques.

Figure 5.9 is a photo of the hologram of a PV cell with no apparent flaws. The ultrasonic and x-ray analyses showed no defects or problem areas. The hologram, however, paints a somewhat different picture. Indeed, on the active part of the cell (in this case considered to be the dark rectangular central part of the cell) the hologram does not show any defects or potential problems except for the bull's eye being centered above the center of the cell. However, look at the interconnects. Several interesting fringe patterns exist on the upper buss bars of the interconnects showing that twisting and pulling actions are taking place that could lead to premature failure of the PV cell or at least possible damage due to the interconnects detaching from the cell. The upper buss bar shows a twisting and rotating action that is causing the problem, especially as these buss bars are so stiff and, therefore, capable of applying significant force to the interconnect legs. Looking at the top of Figure 5.9, the buss bar appears to have two nodes with motion occurring between the nodes and on either side. In addition, judging by the

Figure 5.9. Good cell hologram.

fringe pattern on the buss bars on the left and right sides of Figure 5.9, these interconnects are being rotated by the upper buss bar.

Figure 5.10 provides a closer look at the problem. This figure shows a magnified view of the buss bar and interconnect leg interaction. Notice the fringe pattern on the buss bar first. It shows a rotation in the vertical direction; the buss bar appears to be rotating out of the page in the vertical direction. This action has an interesting effect on the legs that are attached to the buss bar and the cell active area. Notice that the fringes on the legs are vertical (or almost so), implying movement in the left to right direction; the legs look like they are being peeled off the cell. When this hologram was made, the legs were still attached to the cell and no adverse effects could be seen on the active area of the cell, however, it should be clear that several cycles of heating and cooling and stressing the interconnects would lead to problems. Another interesting feature shown in Figure 5.10 is the different number of fringes on the interconnect legs. Obviously, the interplay between the buss bar, the legs, and the attachment points on the cell is complex.

Now consider a photovoltaic cell that has a more obvious defect. Figures 5.11 and 5.12 show a cell that was in relatively good shape initially, but after many thermal cycles developed a crack in the alumina substrate on the back of the cell. Visually, from the front of the cell no difference could be discerned. Upon heating and making a hologram the defect became obvious. The two figures show that the extent of the crack exceeds beyond the active area of the cell in the vertical direction. The two halves of the cell

Figure 5.10. Close-up of interconnects.

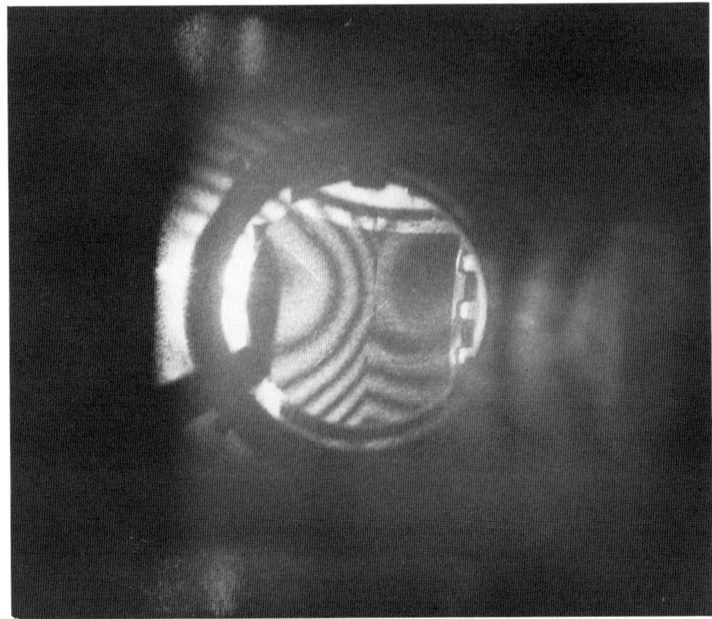

Figure 5.11. Defective cell—view I.

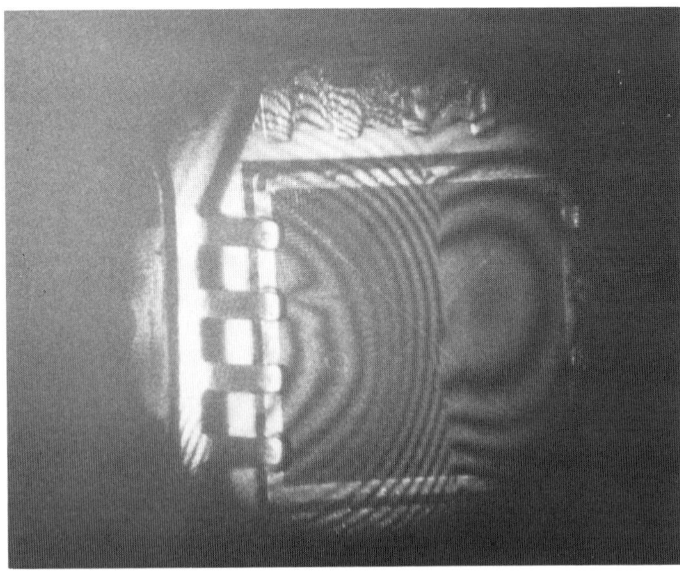

Figure 5.12. Defective cell—view II.

act almost independently of each other showing distinct differences in displacement and origin of the fringe patterns. In Figure 5.11 the fringe hills emanate from the left side moving right, In Figure 5.12 the left side of the cell is clearly displaced more than the right side. The number of fringes is a function of temperature and, apparently, aging. Notice from Figure 5.12 that the interconnect legs on the left side of the cell are detached. This was true when the cell arrived at NMSU from Sandia. In addition, notice the fringe patterns on the right and left buss bars. The density of the fringes is clearly greater on the right side, implying greater relative movement even though the legs are still attached to the cell active area. The fringe patterns still imply a rotating feature to the behavior of the buss bars. There is one other important feature visible in Figures 5.11 and 5.12. When the second interconnect leg from the top on the left side detached from the cell, it left its mark on the active area of the cell. This feature was studied in more detail.

Consider Figure 5.13 which is a photo of the detachment area. The fringe pattern is (sharp as in coming to a point) on a line running from left to right at the point where the leg detached from the cell. Obviously, the detachment from the cell caused severe damage. This occurred well before the alumina substrate cracked. Currently, this region of the cell is being magnified even further to show greater details. It is believed that in the near future the field could undergo a magnification of at least 2 with a better lens arrangement to combat aberrations.

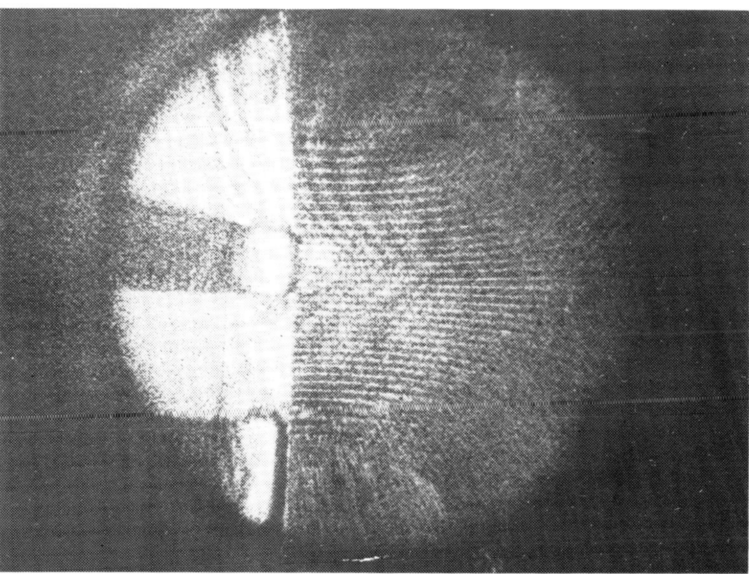

Figure 5.13. Defective cell—close-up.

The study of this cell shows that not only can holometry be used to investigate new or used cells, but also cells that have failed during usage and are of no more practical use for electrical power generation. The mechanisms that caused the failure might be identified in a postusage period and certainly the overall effects of the failure could be studied. One possibility is to record the holometric pattern of several cells before they are used to generate power and then repeat the process for any cell that subsequently fails during operation. A long-range testing program could also be set up to monitor cells that are still operational by repeating the holometric technique over a period of approximately 6 months until the cell fails.

Cell assemblies with the addition of thermal grease placed between the alumina substrate and the aluminum assembly mount were thermally stressed to determine the effect of the grease. The real-time holograms made of these cells did not display a bull's eye fringe pattern as expected. This result is being further researched.

Summary and the future

As can be seen from the discussion of Figures 5.8 through 5.13, holometry can indeed be used as an important complimentary tool to x-rays and ultrasonics. Detailed internal structure of the PV cells cannot be ascertained from holometry with the same confidence as x-rays or ultrasonics; however, the thermal cycling and real-time testing of PV cells using holometry can in some cases show more details about how internal defects will actually behave during use. Holometry's greatest strength is the ability to study the interconnects and the cell deformations under thermal loading. This is something x-ray and ultrasonic techniques will not be able to do as well, if at all.

The next question to be considered is, "What does the future hold for holometric methods and PV cells?" Given the results of this study, holographic interferometry can and should become an important tool for analyzing potential and actual problem areas in new and old PV cells. Temperature and mounting effects should be studied in more detail to improve cell construction as well as the mounting effects on the long-range performance of the cell. New cells should be characterized and placed into operation with at least semiannual recharacterization to monitor changes. At the same time accelerated testing and use of PV cells can be performed while monitoring the cells holographically.

Holometric analysis of cells that have failed is also an important area that should be initiated and supported. The resulting analysis of the cell deformations can be cataloged and used to improve the design of new cells and their mounting structures. In addition, more attention should be paid to using holometric techniques for studying the interconnects and their role in leading to failures of the cells. More work should be done to increase the magnification of areas around the interconnects attachment points while improving the clarity of the field of view. Also, new holographic cameras that produce holograms quickly, that are stored on film should be considered so that a permanent 3-dimensional picture of the PV cell can be recalled and analyzed at a later date.

6
Off-table holography

LARRYL MATTHEWS and BRUCE HANSCHE

Large tensile/compression testing machines, as well as various multipurpose testing machines, are used to evaluate new materials and new product designs. Traditionally, components tested in these machines have been instrumented with strain gages or simply tested until failure occurs. Changes in component performance due to different loading procedures cannot be investigated completely using strain gages and even the correct placement for strain gages is often in doubt. Holography (only continuous wave systems are considered here) offers a technique for full-field investigation of the test component showing unexpected anomalous regions and regions of high surface displacement. This qualitative information can be used to visualize different component behavior under different loading situations and it can also show regions where strain gages should be placed to obtain more detailed data. Quantitative data from holographic interferometry (holometry) is not so readily available. Some analysis, however, is usually possible. As mentioned, only continuous wave (CW) systems are discussed here. Continuous wave systems are typically less expensive, less bulky, and easier to transport and use in multiple locations than pulsed systems. Also, real time holograms are more readily available from CW systems.

Making holograms of objects mounted in large testing machines is not an easy task. The testing machine is not isolated from vibrations originating from other locations and in addition tends to generate significant vibrations of its own. Making a hologram in this environment is difficult in the least and impossible under certain circumstances. Success depends on the actual test environment, the holographic equipment, and the procedures used. This chapter is concerned with explaining how successful off-table holograms can be made and the potential pitfalls that await the experimenter.

Experimental set-up

Methods to record successfully holograms of objects that undergo motions relative to the holographic medium are outlined in Erf (1). Pulsed laser, speckle reference beam, and local reference beam generation are techniques often mentioned. Local reference beam generation is the technique discussed in this chapter. Erf describes the basic set-up but does not go into any of the actual details or potential problems of the procedure. Figure 6.1 is a sketch of the set-up used for one of the tests conducted at the Sandia National Laboratories Static Testing Lab. Figures 6.2, 6.3, and 6.4 are photos of the actual test object and two equipment set-ups.

Historically, Corcoran, Herron, and Jaramillo (2) were the first to describe the procedure in 1966. Basically, a mirror is attached either to the object or to the testing machine that houses the object whereupon part of the object beam is intercepted and redirected to the holographic plate, thus generating the reference beam. One of the advantages of this procedure is the simplicity of using single beam illumination. One of the problems is adjusting the position of the object beam and reference mirror so that the correct reference to object beam ratio can be obtained.

The reference mirror has been attached to large objects, using dental cement, allowing good real-time and double exposure holograms to be made. Helium–neon lasers of 50 and 15 mW have been used to illuminate the object. This is convenient because of the continuous wave operation of these lasers and the ease of use and availability. When the objects were small, the mirror was attached to the testing machine using magnets, as shown in Figure 6.2.

Theoretically, this procedure is supposed to work by having the mirror move with the test object and thus compensate for any unwanted motions introduced into the system by machine vibrations. In practice, the mirrors do move with the object, however, the useful illuminated region of the object is usually smaller than can be achieved with a completely isolated system. The region on the object nearest the mirror always has the best fringe contrast with the contrast decreasing as you move away from the mirror. A revised arrangement, Figure 6.4, was used to greater success. The basic principal used was to attach the holographic system to the testing apparatus so that the two moved together. This showed marked improvement in the fringe field.

Experimental procedure

Because of the nature of off-table holography, it was evident that several holograms would have to be made to successfully track the real-time fringe build-up during loading and unloading of a test object. For this reason automatic holographic cameras were investigated to determine the best choice for the problems at hand.

The criteria used to choose the camera included,

1. Ease of operation.
2. Flexibility
3. In situ exposure and development

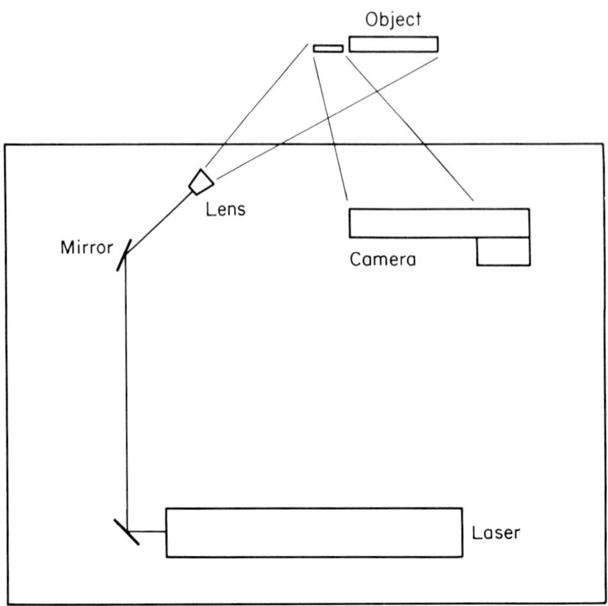

Figure 6.1. Sketch of the set-up used for one of the tests conducted at the Sandia National Laboratories Static Testing Lab.

Figure 6.2. Object under test with off-table holography.

Figure 6.3. Object under test with off-table holography.

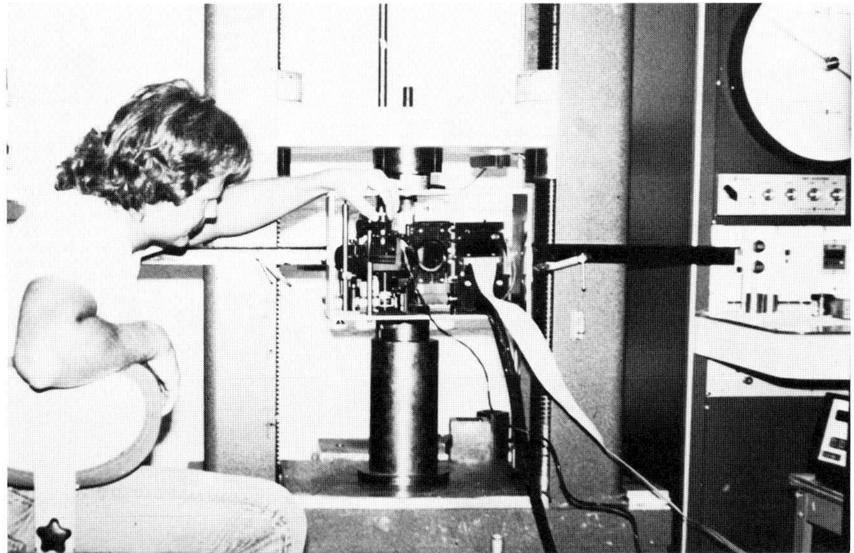

Figure 6.4. Object under test with off-table holography.

4. Quick operation
5. Ability to record down to dc ranges
6. Portability

Several automatic camera systems are available on the market and could be used for this problem; however, the LTI holocamera 1800 (Laser Technology Incorporated, PA) and the Keystone Scientific 70-mm systems (Keystone Scientific Co.; Thorndale, PA) were chosen. These camera systems used conventional holographic emulsions on a 35-mm and 70-mm roll format. The camera houses the film in a plastic enclosure with water being injected in the enclosure directly behind the film. This holds the film in place for exposure and viewing. When the film is exposed the water is displaced by developer for approximately 30 seconds where upon water is injected again and the hologram is ready for viewing. When another exposure is desired the camera advances the film to the next unexposed segment and the exposure and develop sequence is repeated. Only two buttons must be depressed to execute the sequence. Total time for this procedure is approximately 45 seconds. The need for dc capability precluded using thermoplastic systems because of the limited bandwidth of the median.

As shown in Figure 6.1, the LTI holocamera was positioned on a table along with the helium–neon laser, mirrors, and lenses. This table was rolled up to the Tenius-Olsen universal testing machine and positioned as close to

Figure 6.5. Representative test.

Figure 6.6. Test set-up for large object.

Figure 6.7. Large object under test.

the test item as possible. Figure 6.3 shows the test apparatus and laser/holocamera table.

The laser illuminated the test item and the reference mirror (Figure 6.2) in preparation to make a hologram. The 15-mW laser was used with exposure times of approximately 1 second, with reference to object beam ratios of approximately 9 to 1. In most cases acceptable holograms were made; however, in some instances the holocamera and the test item had to be coupled together with an aluminum plate.

Figure 6.5 is a photograph of a representative test. The reference mirror is clearly in the field of view with the test item and the resulting real-time fringes are on the left. Note that no motion fringes are evident on the reference mirror support. Real-time fringes were relatively easy to acquire and follow over a considerable loading range. Initially, loading caused only linear horizontal fringes on the test item. Approximately 500 pounds of force was applied until too many fringes were on the test item to be seen. The load was increased and another hologram was made. This procedure continued until the test item began to show some plastic deformation, evident from the curved fringe pattern.

The procedure described was attempted with success on two other testing devices at Sandia. A space frame used at Sandia for testing large objects and applying lateral loads was tested next. The test set-up and the test item are shown in Figures 6.6 and 6.7. Because the holocamera and the test item were tied together, instead of on the Tenius-Olsen universal testing machine, the results were quite satisfactory. A larger Tenius-Olsen universal testing machine was used next; however, the testing procedure had to be modified to accommodate the different environment. The larger Tenius-Olsen had a false floor around it and more vibrations were generated by using this machine. This required attaching a damped table to the testing machine and attaching the reference mirror to the test object. No problems arose from this procedure as the test item used on this machine was large and the operation of the testing machine was unhindered. After these activities, the Keystone camera was used in the box attachment shown in Figures 6.8 and 6.9. The stability of the fringes, the area over the object upon which the fringes developed, and the ease of making the hologram make this system very useful.

The procedures described in this chapter were found to be useful to produce real-time off-table holograms to evaluate test items that could not be accommodated on a vibration isolated system. The basic idea was object-coupled reference beam generation that allowed the reference mirror to follow random motions (due to extraneous vibrations) of the object so that the reference object beams deviated by less than one-half of a wavelength during generation of the initial hologram. Greater relative fringe movement than this would not allow the initial hologram to be made. It became evident during the course of using the first set-up (Fig. 6.3) that the best fringes were found on the object in the immediate vicinity of the

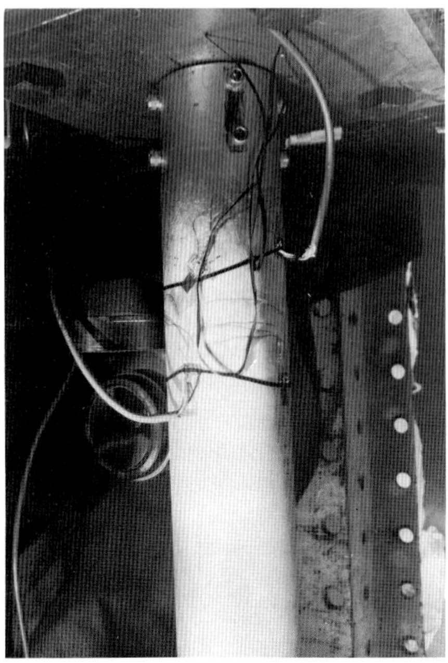

Figure 6.8. Box attachment for Keystone Camera.

Figure 6.10. Optical fiber in place for reference beam.

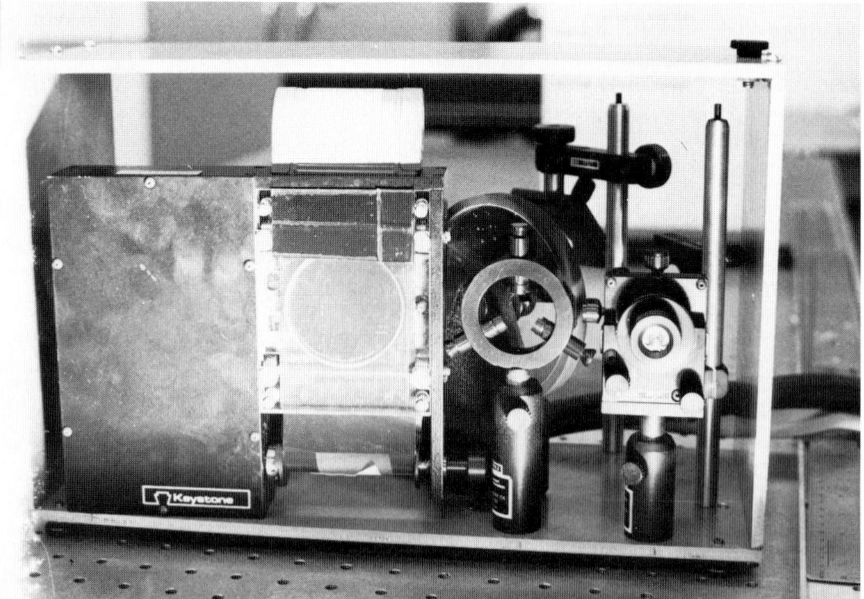

Figure 6.9. Box attachment for Keystone Camera.

reference mirror with the quality of the fringes falling off rapidly away from the mirror. In an effort to improve the procedure an optical fiber was attached to the object for the purpose of providing the reference beam (Fig. 6.10).

This attempt was not successful and after several tries and different configurations an analysis of the system proved that the fiber reference beam would not be successful. A short explanation of the analysis follows. Once the optical fiber is attached to the test object it moves easily as the test object moves. As the object moves, however, the object beam can travel a distance up to twice this distance. This can be seen by visualizing the object beam first traveling the extra distance to the object (assuming the object moves away from the laser), then traversing the same distance again to reach the holocamera, whereas the optical fiber traveled with the object increasing its path length by only half that of the object beam. Therefore, the optical fiber reference beam cannot track the movement of the object correctly for a hologram to be made.

Conclusions

In this chapter techniques for performing off-table holometry have been described. Successful holograms have been made that have been used to analyze qualitatively test objects under load. Although double exposure holograms were made from the set-up in Figure 6.1, construction of real-time holograms was the main objective. Real-time holograms offer dynamic information about the test object as it is happening and over a continuum of time. In addition to the real-time information, another objective was to use continuous wave helium–neon lasers that are readily available and easy to operate.

The essential things to remember are to generate the reference beam from the object location and tie the object and the holocamera together whenever possible. Also remember to choose a holocamera that can record down to dc as the angle between the reference beam and the object beam is usually small (certainly less than 20 degrees). It is also necessary to dampen the table holding the laser and the holocamera, especially if it is attached to the testing machine. The only unsuccessful part of the effort involved the use of optical fibers for reference beam generation.

In conclusion, off-table holometry is possible and should be used for full-field analysis of test objects, as well as the analysis of loading mechanisms and proper strain gage placement.

Acknowledgment

The work described in this chapter was performed at Sandia National Laboratories, Organization 7551, Albuquerque, New Mexico.

References

1. Erf, R. K. *Holographic Nondestructive Testing.* 1974, Academic Press.
2. Corcoran, V., Herron, R., Jr., and Jaramillo, J. *Appl. Opt.* **5,** 668 (1966).

7
High temperature optical strain measurement

KARL A. STETSON

The measurement of strain by optical techniques has been pursued for primarily three reasons. First, many optical techniques provide the ability to measure object displacements over a wide field of view. Unlike discrete strain gages that provide measurements only at predetermined locations, optical techniques make it possible to determine where on the object the maximum displacements occur. Second, some optical techniques are non-invasive and require no preparation or alteration of the object surface or any physical contact. Third, optical techniques offer the possibility of performing metrology at higher temperatures than are practical for electrical strain gages.

Of the three reasons cited, only the last one has proven to be compelling. The advantage of full-field measurement is largely illusionary, because most often the peak strains on a structure occur inside bolt holes or inside mounting slots where optical access is not available. Furthermore, finite-element modeling can solve the problem of locating peak stresses, provided the models have been validated by strain gage measurements at representative locations. The advantage of noninvasive measurements may not be critical for many robust structures such as jet engine disks or airframes. Conventional strain gages, however, are ill-suited for measurements at extremely high temperatures, and this leaves the designer with unvalidated finite-element modeling beyond a few hundred degrees Celsius. It is primarily for this reason that optical techniques of strain measurement are pursued: to allow validation of finite-element modeling at high temperatures by making representative measurements at visible locations.

Optical measurement techniques

The following is a brief description of the various optical methods that have been used to measure strain.

Photoelasticity and moiré interferometry

Before 1964, there were essentially two techniques available for optical measurement of strain and displacement, photoelasticity and moiré interferometry. Photoelasticity is an excellent method for strain measurement that is based on modeling the structure in a transparent solid that exhibits stress birefringence. Loads applied to the structure distort its molecular structure sufficiently to alter the state of polarized light, and the complete stress state of the material can be determined by a number of tests. Moiré interferometry is accomplished by marking a grid of fine lines on the structure. Superimposing photographs of the grid on the structure, before and after loading, can reveal displacements on the order of the grid spacing as broad interference fringes.

Hologram interferometry

It was discovered in 1964 that interference effects could be observed between the fields reconstructed by holograms that had double or multiple exposures, and that interference could be observed between the field reflected by an object and its hologram reconstruction. The ease with which hologram interferometry allowed visualization of object deformations launched a revolution in optical metrology. The complexity of interpreting the fringe patterns, however, and in particular the problems of fringe localization, delayed the use of holography in strain analysis for nearly 10 years. Nonetheless, numerous techniques for strain analysis have been developed using hologram interferometry.

Laser speckle techniques

Interferometry and metrology based on the use of laser speckles grew out of work in hologram interferometry and dates from 1969. Two phenomena are used in ways that are akin to moiré techniques. First, when two randomly speckled fields interfere, the resulting speckle field displays a cyclic correlation with itself every time the phase difference between the fields is changed by 360 degrees. The correlation changes can be converted to fringes by double exposures or superimposition of photographs of the resulting speckle fields. If the phase differences are generated by an object deformation, the fringes can be used to measure it. The second phenomenon is that the speckles observed on an object appear to move as if attached to the object. As such, they can provide a random moiré pattern of high constrast and fine texture that need not be resolvable by the camera lens. Methods have been devised to extract object displacements and strains from double-exposure and single-exposure speckle photographs.

Laser extensometry

A number of laser extensometers have been developed that can be used for strain measurement. One of the simplest is a system developed by Zygo Corp. that sweeps a beam of laser light across a field of view at a constant speed without angular deviation. The sample under load is configured to have shoulders that obstruct the beam at the ends of its excursion. The time for which the beam is not obscured is a measure of the length of the sample.

Another extensometer, developed by W. N. Sharpe (1), employs either small grooves or small indentations in the surface of the object. These reflect light from a laser beam into an interference pattern that changes in response to surface strain. Measurement of these changes can be accurately correlated to strain.

A similar technique was introduced by I. Yamaguchi (2) that employed the light scattered by the untreated object surface. The object was alternately illuminated by two laser beams striking the surface at equal and opposite angles to the surface normal, along which it was observed. Speckle displacements are measured and correlated to surface strain.

Another system was developed by K. A. Stetson and R. K. Erf (3) that illuminated two nearby points on an object surface, along the normal, and observed the scattered light from equal and opposite angles. The observed light is modulated at Fourier transform planes by photographic recordings made with the object unloaded. A Doppler shift is introduced between the two beams that illuminate the object and strain is recorded as the phase change of the fluctuations of the observed light beams.

Finally, a recent development by M. Hercher, G. Wyntjes, and H. DeWeerd (4), of Optra, Inc., combines features of the other systems. Here the object is illuminated at one location by two beams, at equal and opposite angles to the surface normal, which are Doppler-shifted. The light observed along the surface normal fluctuates at the Doppler frequency and the phase of these fluctuations changes in proportion to the displacement of the object under the point illuminated. A pair of such sensors can be used to measure strain.

Applicability to high-temperature strain measurement

The following is a discussion of the applicability of the various methods of optical strain measurement to high-temperature environments.

Photoelasticity

Photoelasticity is not a technique that has been used for high-temperature strain measurement. The main reason is that transparent materials are not available that can withstand high temperatures, exhibit good stress birefringence, and be easily machines. Also, for most applications, the testing must be done with the real material under study and not with another material whose characteristics may not match.

Moiré interferometry

Moiré interferometry has been used for strain measurement at extremely high temperatures (5–8). Grid patterns of up to 40 lines/mm have been fabricated on refractory materials that are durable up to temperatures of 1370°C (2500°F). They are formed by etching through patterns created in photoresist or by coating the object surface with refractory paint and pressing a wire grid into the paint. The grid patterns may be photographed at various times during a test sequence and pairs of photographs can be compared later to evaluate strain.

To obtain strain measurements, the photographs are superimposed and observed. If the pattern on one photograph is in phase with that on the other, constructive moiré interference is observed, and if they are out of phase, destructive interference is observed. The distance between the centers of bright or dark fringes is the distance separating points on the object that have moved relative to one another by the pitch of the photographed grid. Strain, therefore, equals the grid pitch divided by the moiré fringe spacing, which becomes the gage length of the measurement.

$$\varepsilon = g/m, \tag{7.1}$$

where ε is strain, g is the grid spacing, and m is the spacing of the moiré fringe.

It is clear from Equation 7.1 that high-strain sensitivity and short-gage lengths are mutually exclusive in this technique. Short-gage lengths occur at locations of high strain. For example, for a gage length of 1 mm and a grid of 40 lines/cm, the minimum measurable strain would be 25,000 microstrain. With the same grid, a gage length of 25 mm would be required to obtain a strain resolution of 1000 microstrain. To obtain a strain resolution of 100 microstrain at a gage length of 1 mm would require a multiplication of sensitivity by 250 times. Multiplications of 20 to 30 have been obtained with photographic grids of high quality (9) by observation of the gratings by their higher diffraction orders. It has not been established that such high multiplication could be obtained with gratings photographed on objects at high temperatures.

A heterodyne technique could be applied to the readout of moiré grids that might allow an easy increase in sensitivity and an increase in measurement range. If a single photograph of the object grid pattern were illuminated by a two-beam interference pattern of matching spatial frequency it would exhibit moiré interference. If a Doppler shift were introduced between the two interfering beams, the moiré interference fringes could be made to scan across the field of view and would generate a sinusoidal irradiance fluctuation at every location. A pair of photodetectors located beyond the photograph would generate a pair of output signals whose phase difference would be a measure of the mismatch between the interference pattern and the photographic grid. The difference between the phase measured for one photograph and the phase measured for a second photograph would be a measure of the strain exhibited between the two photographs. Phase-stepping techniques have been developed that allow rapid calculation of phase distributions by means of diode array TV cameras, which could be applied to this problem (10). The accuracy that could be obtained by this technique would vary from 25 to 250 microstrain at a gage length of 1 mm depending on the precision of the phase measuring technology. This would also have the advantage that, because the fringes are moved across the field of view, strain data could be obtained anywhere on the photographs and not just between fringe centers.

In the final analysis, however, there is a disadvantage to the use of moiré interferometry in that it requires preparation of the object surface. Such preparation may affect such parameters as heat transfer from the surface, which may be an important aspect of high temperature testing. Also the camera must be focused accurately on all parts of the object surface so that the grid can be photographed. This presents difficulties when working with surfaces that are curved.

Hologram interferometry

There have been very few publications to date on applications of hologram interferometry to objects at elevated temperatures. The numerous difficulties presented by high temperatures require the use of either pulsed lasers (11) or electronic speckle pattern interferometer (ESPI) systems (12). The work reported has been limited to recording vibration patterns on panels (11) and to the study of phase changes and the growth of oxide layers (12).

Although numerous techniques are available for holographic strain measurement (13), none of these have been applied to objects at high temperatures, nor are they well suited to high temperature environments. The techniques available are all equivalent to determining vectorial displacements at nearby object points and subtracting these values. These require, therefore, multiple illuminations or observations of the object. If the gas through which these light beams must pass is not homogeneous, spurious fringe patterns are introduced in the hologram reconstructions and these lead to serious errors. Because the holographic plates and the ancillary apparatus must be kept at reasonable temperatures, and there must be some material in the form of gases or windows separating them from the object, it is very difficult to achieve a homogeneous material through which to illuminate or view the object.

Laser speckle techniques

Laser speckle techniques have been applied to high-temperature strain measurement in the form of speckle photogrammetry (14). In this technique, single-exposure photographs are recorded of the object under study with laser illumination. These photographs, called specklegrams, are analyzed in a heterodyne interferometer. In this instrument, the specklegrams are placed side-by-side on positioning stages that allow adjustment so that common areas on each are illuminated by narrow, converging, mutually coherent laser beams. The diffraction halos from the two specklegrams are combined by means of mirrors and a beamsplitter to obtain an interference pattern that can be made to scan by Doppler shifting the two illumination beams. Photodetectors in the output field convert the fluctuating irradiance to electrical signals. The phase of the output signals changes in proportion to strain as the beams illuminating the specklegrams are scanned vertically and horizontally.

Although numerous methods of strain measurement have been reported using laser speckles (15), heterodyne photogrammetry is the only technique that has been successfully demonstrated for strain measurement at high temperatures. The most common form of strain measurement using speckle correlation uses two beams to illuminate the object from equal and opposite angles to the observation direction. Lateral translations of the object cause cyclic variations in correlation between the speckle fields observed due to each illumination, and double-exposure photographs can be processed to detect the correlation variations as fringes. This geometry, like those used for hologram interferometry, allows inhomogeneities in the hot gas surrounding the object to introduce errors by introducing patterns of phase difference between the two illumination beams. Such patterns show up in the output photographs as spurious patterns of apparent strain.

Speckle photography relies only on the displacement of the speckles formed in the image of the object and is nearly immune to phase patterns in the illumination beam. The most commonly used technique employs double-exposure photographs from which image displacements are extracted by a technique called halo fringe analysis. A small, narrow, converging beam of monochromatic light is passed through an area of the photograph and the doubling of the speckle structure in the photograph causes fringes to appear in the halo of light diffracted away from the zero order. Analysis of the spacing and angle of the fringes allows determination of the average vectorial displacement between the speckles of the two exposures. The main disadvantage of this technique is that its range of measurement is limited to little more than a factor of ten for any single optical set-up. This is very restrictive because the technique is dependent on having the object displacements lie within those limits to make measurements. Also, a high numerical aperture is required of the optical system to obtain the high displacement resolution needed for accurate strain measurement. If the object moves in the axial direction by more than the focal depth of the lens system, the speckle patterns decorrelate and measurements of displacement become impossible. Because the depth of focus of a lens system decreases inversely with the square of the numerical aperture, it is convenient to avoid high numerical apertures. Finally, axial displacement of the object creates apparent strain by changing its effective magnification if the lens system has a spherical perspective.

Heterodyne speckle photogrammetry provides a solution to these problems. First, specklegrams are recorded as single exposures. Lateral object displacements of 5 mm or more can be tolerated quite easily by proper alignment of the pair of photographs in the photocomparator. Low numerical apertures can be used (corresponding to $f/10$ or $f/20$) because the heterodyne readout technique for the halo fringes is 10 to 100 times more accurate than other methods of fringe measurement. This, in turn, makes it practical to design telecentric lens systems that can remove apparent strains due to axial displacement of the object. Theoretical strain resolution for this technique is in the order of 10 microstrain for a 1-mm gage length. Practical results, however, are limited by fringe contrast to the order of 100 microstrain. The strain range is virtually unlimited, provided sufficiently gradual steps can be applied.

There have been mainly two problems encountered with speckle photogrammetry in practical applications up to 870°C. The first was image blurring due to gas flow at high pressure and high temperature (16). It was determined that, in a combustor test rig, pressures greater than 3 atmospheres caused serious degradation of specklegram data and created patterns of apparent strain that were random and in the order of 2000 to 3000 microstrain. The second problem was loss of speckle correlation due to tilting and defocus of an object in a cyclic fatigue rig (17). This problem was encountered with a 3-mm thick sample of jet engine burner liner material that was heated from the rear while held against a frame. The sample became permanently warped by more than 10 mm. Speckle photogrammetry is applicable to the measurement of strain on structures at temperatures up to 1370°C provided proper care is taken with surface stability and turbulance in the gas surrounding the object can be controlled.

Laser extensometry

All of the laser extensometry systems cited in the introduction have been used to measure strain at high temperatures. The following is a description of the advantages and disadvantages of each method.

The Zygo system

The Zygo system has been used for high-temperature strain measurement in a variety of applications. The technique has the advantage of being relatively immune to problems of thermal gradients in the gas surrounding the object. It has the disadvantage that the object must be specially configured to apply the technique, and the gage length is fixed by the geometry of the sample. Furthermore, the strain measured is the average over an extended region. Finally, some problems have been observed with the update rates (in the order of 10 readings/sec) being too slow to measure relaxation phenomena.

The Sharpe method

The Sharpe method has been used up to temperatures of 730°C (18) and may be used to higher temperatures. It can be applied to ceramic materials by cementing platinum tabs to the structure into which the reflective indentations can be made. A wide variety of readout techniques can be used in this system to give a variety of response speeds ranging from 0.25 sec to 10 microseconds. The gage length can be small, in the order of 100 micrometers, and the range of measurement can exceed several percent. The sensitivity can be adjusted by varying the spacing between the reflective indentations and can range down to 200 microstrain. The disadvantages of the system include the fact that displacements along the surface normal will generate apparent strains in the readout. The system also requires large angles (about 90 degrees) included between the observed reflections, and this makes the measurements subject to perturbation by hot turbulent gases surrounding the object, and this problem does not appear to have been studied.

The Yamaguchi Method

The Yamaguchi method, initially developed by Yamaguchi, has advantages over the system developed by Sharpe. Aside from the obvious advantage of working with the untreated object surface, the method of illuminating the surface from equal and opposite angles to the surface normal eliminates apparent strains due to displacement in that direction. The disadvantage is, however, that the illumination must be switched from one direction to the other before the load is applied to the object and once again after the load is applied. The four speckle patterns observed along the surface normal must be stored and processed to measure the speckle displacements to obtain strain outputs. This process requires in the order of 10 seconds, which is more time than may be desirable for many applications. Nonetheless, the strain sensitivity has been reported in the order of 20 microstrain with a range that is essentially unlimited.

Desire for more rapid response had led Yamaguchi to develop the alternate configuration of illumination along the surface normal and observation from equal and opposite angles (19). Although this set-up is subject to apparent strain readings due to displacement along the surface normal, it has the capability of generating strain data at rates exceeding 5 kHz. This is accomplished by an electronically scanned detector array that samples the observed speckle patterns. In both configurations, however, the angle subtended by the measurement system is approximately 90 degrees so that turbulent gases surrounding the object have the potential for creating errors. The effect of this problem has not been studied in detail.

The URTC System

The strain sensor developed by K. A. Stetson at United Technologies Research Center (URTC) has similarities to those of Sharpe and Yamaguchi but eliminates some of the problems associated with those systems. The configuration of illumination along the surface normal and observation at equal and opposite angles would allow out-of-plane displacements to generate apparent strains. This effect is eliminated, however, by having a Fourier transform lens in each observation channel. The use of optical heterodyning increases the readout accuracy to the point that the angle subtended can be reduced to 8 degrees and still yield 10-microstrain resolution. This lowers the sensitivity of the system to perturbations by hot gases. The update rate is dependent on the heterodyne frequency that can be set to a few hundred kilohertz.

The main disadvantage with the system is the use of a photographic recording to modulate the light passing through the Fourier transform plane. This has the effect of recording the initial state of the object and provides a reference against which the deformed state can be compared. As the object undergoes deformation it may move out from under the illumination spots so that new regions are illuminated, or the surface may tilt so that different fields enter the observation channels. Either of these effects causes the photographic recording to decorrelate with the field patterns irradiating them, and the result is a loss of signal. Even with in situ processing the recycle time for a new recording is in the order of 30 seconds, and this greatly hampers the system in many applications. A newer system of this type is under development at URTC, which will employ an electro-optic readout system to replace the photographic film. This system may allow data rates up to 100 readings/sec. Also the detection scheme used on the Optra, Inc. system, described in the next section, could be applied to this system.

The Optra system

The extensometer developed by Optra, Inc. combines features of the initial Yamaguchi system and the UTRC system. It uses the configuration of illumination at equal and opposite angles and observation along the surface normal that makes it immune to apparent strain from out-of-plane displacements. The update rate is nominally 20 kHz. The basic unit of the system is a single displacement sensor that measures object displacement under the point where the illumination beams cross. To measure strain, however, the system has to be doubled so that four illumination beams are required to measure strain. The advantage of the Optra system over the UTRC system is that no photographic recording is used. Instead, small spots on the object are illuminated and the scattered light is collected over a small area. Analysis of the statistics of this detection process shows that a measurement uncertainty, called a dead reckoning error, accrues each time the object moves out from under the spot where it was previously illuminated. The magnitude of this uncertainty decreases, however, relative to the actual measured displacement so that the percentage error actually decreases. The dead reckoning error can accrue from tilting the object surface from translations at right angles to the direction of measurement, and from corrosion or erosion of the object surface itself. The chief disadvantage of the Optra system is that there is no independent method for assessing the magnitude of the dead reckoning error. The system provides a simple reading that may or may not contain a significant error. The addition of an electro-optic system to keep track of the number of decorrelations of the observed field pattern could solve this problem.

Discussion

A variety of optical techniques are available for strain measurement, and the choice of technique depends on the requirements of the testing to be done. Optical strain testing may be divided into two categories: tests where it is essential to obtain spatial distributions of strain at selected time intervals and tests where strain as a function of time is required at selected object locations. There are, at present, no optical methods that can provide full-field strain measurements at rapid framing rates. In any application of optical strain measurement, therefore, the first decision to be made is whether the strain variations of interest are spatial or temporal. If they are spatial, then techniques, such as moiré interferometry or speckle photogrammetry, can be applied. If preparation of the object surface is not allowable, then speckle photogrammetry is the only viable technique. If temporal strain variations are most important, then there are five techniques of laser extensometry that may be applied. Again, if surface preparations are not allowed, there remain three techniques that are applicable.

Summary

Strain measurements may be divided into two categories: measurement of spatial strain distributions and measurement of temporal strain histories.

1. Spatial Strain Distributions

    A. Moiré Interferometry
Applicability: To structures whose surface can be prepared and to tests where strains greater than 1% are expected.

Strain resolution: Nominally 1000 microstrain but extendable by the introduction of heterodyne readout techniques.

Gage length: Variable and inversely proportional to strain level, however, subject to reduction to a fixed length by heterodyne readout.

    B. Speckle Photogrammetry
Applicability: To structures where high-pressure gas turbulence and surface warping can be controlled.

Strain resolution: 100 microstrain has demonstrated in practice with theoretical limits in the order of 10 microstrain.

Gage length: Nominally 1 mm.

2. Temporal Strain Histories

    A. The Zygo System
Applicability: To extensometry of test bars that can be properly configured.

Strain resolution: 10 microstrains or less depending on sample length.

Gage length: Set by sample configuration, generally more than 25 mm.

Update rate: Typically 10 readings/sec.

    B. The Sharpe and Yamaguchi Methods
Applicability: To set-ups where wide angular subtense is acceptable.

Strain resolution: 20 to 200 microstrain depending on readout techniques.

Gage length: 0.1 to 1 mm

Update rate: Variable from 0.1 reading/sec to greater than 5000 reading/sec depending on readout technology.

    C. The UTRC and Optra Systems
Applicability: To set-ups where narrow angular subtense is required.

Strain resolution: Better than 20 microstrain and dependent on readout technology.

Update rate: 20,000 readings/sec or greater depending on heterodyne frequency.

References

1. Sharpe, W. N. *International Journal of Nondestructive Testing* **3,** 56–76 (1971).
2. Yamaguchi, I. *J. Phys. E: Sci. Inst.* **14,** 1270–1273 (1981).
3. Stetson, K. A., and Erf, R. K. A optical heterodyne strain sensor. Report No. F33615-80-C-2067, AFAPL, Wright-Patterson AFB, Ohio, Sept. 1981.
4. Hercher, M., Wyntjes, G., and DeWeerd, H. *Proc. SPIE* **746,** 185–191 (1987).
5. Sciammarella, C. A., and Rao, M. P. K. *Experimental Mechanics* **19,** 389–398 (1979).
6. Cloud, G., Radke, R., and Peiffer, J. *Experimental Mechanics* **19,** 19N–21N (1979).
7. Bayer, M., and Cloud, G., Moiré to 1370°C. Proc. SEM Spring Conf. on Exp. Mech., New Orleans, LA, June 1986.
8. Cloud, G., and Bayer, M. High temperature moiré. Proc. SEM 4th Annual Hostile Environments and High Temperature Measurements Conf., Windsor Locks, CT, March 1987.
9. Post, D. *Appl. Opt.* **6,** 1938–1942 (1967).
10. Bruning, J. H. Fringe scanning interferometers. In *Optical Shop Testing* (Malacara, D., ed.). Wiley, New York, 1978.
11. Evensen, D. A., Aprahamian, R., and Everoye, K. R. Pulsed differential holographic measurements of vibration modes of high temperature panels. NASA Contract Report 2028, (1972).
12. Løkberg, O. J., Malmo, J. T., and Slettemoen, G. A. *Appl. Opt.* **24,** 3167–3172 (1985).
13. Vest, C. M. *Holographic Interferometry,* Wiley, New York, Chap. 4 (1979).
14. Stetson, K. A. *Proc. SPIE* **370,** 46–55 (1983).
15. Ennos, A. E. In *Progress in Optics,* (E. Wolf, ed.) North-Holland, Amsterdam, Vol. XVI, Chap. 4 (1978).
16. Stetson, K. A. Demonstration test of burner liner strain measuring systems. NASA Report No. CR-174743 (1984).
17. Stetson, K. A. Demonstration of laser speckle system on burner liner cycle fatigue rig. NASA Report No. CR-179509 (1986).
18. Sharpe, W. N. *Opt. Eng.* **21,** 483–488 (1982).
19. Yamaguchi, I., Furukawa, T., Ueda, T., and Ogita, E. *Opt. Eng.* **25,** 671–676 (1986).

III
Industrial inspection and documentation

8
Holographic documentation and inspection

LLOYD HUFF

During the late 1970s and early 1980s, the University of Dayton Research Institute (UDRI), under contract to the Air Force Weapons Laboratory, provided a variety of support to the development of optical component technology for the Air Force's high-energy laser program. One of the tasks on this contract was a test program to evaluate the performance of new optical component materials (both substrates and coatings) under exposure to various levels of high-energy laser irradiation. Determination of damage thresholds was of particular interest. Small samples of approximately 1-1/2 inches in diameter were tested. Meaningful reduction of the test data and accurate evaluation of the performance required careful pretest characterization; recording the surface microstructure of the samples was especially important. This was typically done by taking a series of high-magnification photomicrographs, an expensive and time-consuming task.

Our Air Force Project Officer suggested the use of holography to reduce the pretest characterization effort. The concept was to replace the series of photomicrographs with a single high-resolution hologram of the test piece. The holographic image could be examined microscopically at any desired level of magnification and the pretest conditions of the sample could be compared with posttest morphology (including damage sites), after the sample underwent high-energy laser exposure. Consequently, the holographic documentation system was constructed for this purpose. The results were indeed satisfactory and the system was successfully used to replace the tedious method of photomicroscopy (1).

During the evaluation and successful application of this documentation system we recognized the potential for applying the technology to other areas. The possibilities were discussed with workers in the integrated circuit field and funding was obtained to apply the holographic documentation system to defect detection in integrated circuit photomasks and wafers. During this effort a new holographic optical processing technique was developed that made possible the extremely successful application of classic Fourier transform spatial filtering techniques to submicrometer defect detection (2–5).

Holographic documentation

Adequate pretest characterization of high energy laser optical component test samples required better than 10-μm image resolution. A holographic system designed to meet or exceed this resolution with a 2-inch diameter working field was developed. A simple and straightforward approach to apply holography to this problem is simply to make a far field hologram of the test piece and reconstruct the real aerial image of the sample with the conjugate reference beam. Microscopic inspection of the aerial image makes possible examination of both surface and volume details of the sample as the microscope objective can probe through the depth of the aerial image space. The first attempts to solve the documentation problem with this simplified approach encountered difficulties in both limited resolution and severe background noise. Higher resolutions and lower noise were observed with image plane holograms wherein the test piece was imaged onto or near

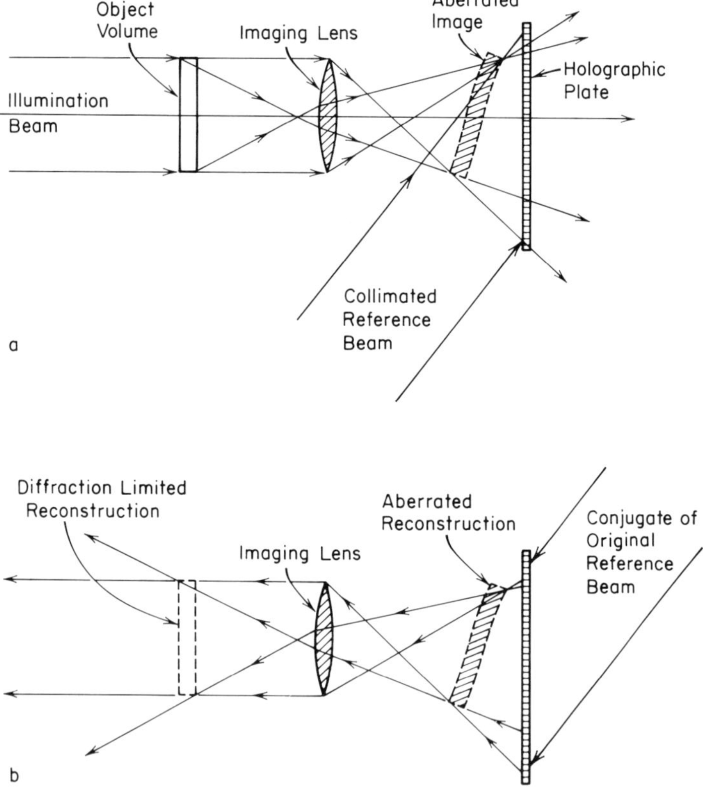

Figure 8.1. Holographic recording (a) and reconstruction (b) of aberrated image formed by lens. b. Shows diffraction-limited real image formed due to reverse ray tracing through original lens.

the hologram film plane. The lens used for this purpose does not need to be diffraction-limited as aberrations are compensated by the reverse ray-trace reconstruction beam. This property is illustrated in Figure 8.1.

Several types of optical systems ranging from a simple single lens to complicated telescope systems were investigated for imaging the test piece onto the hologram film. Resolution of the final image is limited by the effective f/number of the system, which is in turn limited by both aberration and vignetting. We observed the best performance with a single lens imaging and test piece onto the film with unity magnification.

The final system for documenting the optical component test samples is shown schematically in Figure 8.2. An f/3.2 lens was used and an Argon laser operating at a wavelength of 514.5 nm was the illumination source. A polarizing beam-splitting cube and a half-wave plate were used to split the beam into object and reference beams and to adjust the beam ratios. A second polarizing beam-splitting cube and quarter-wave plate were used to efficiently illuminate the test target and direct the reflected light to the holographic plate. The holographic image is reconstructed with the conjugate reference beam by removing M4. The test target is removed and the holographic image of the sample is examined with the aid of a microscope.

AGFA 8E56 HD plates were used with various developing processes, including the standard D19 processes. Of all the methods studied, PAAP and HRP/bromine methanol provided the highest resolution and lowest noise.

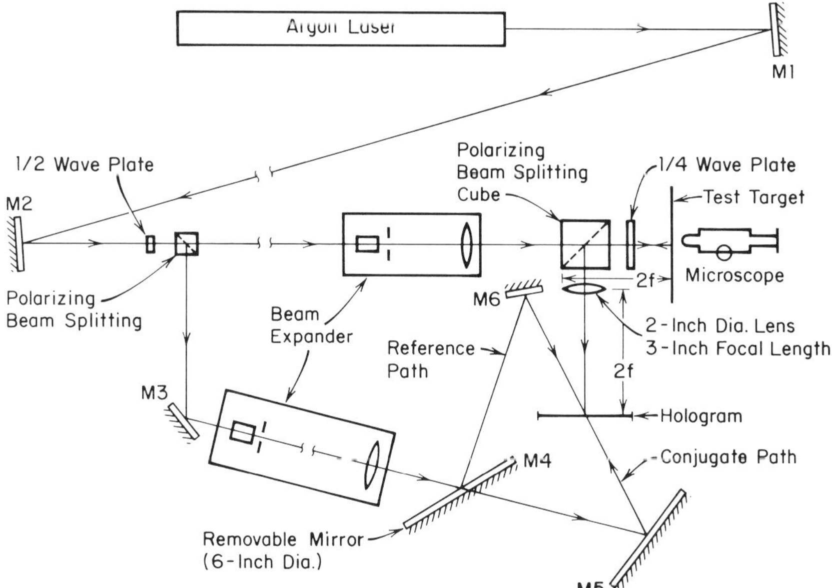

Figure 8.2. Holographic documentation optical system using simple one-to-one imaging optics.

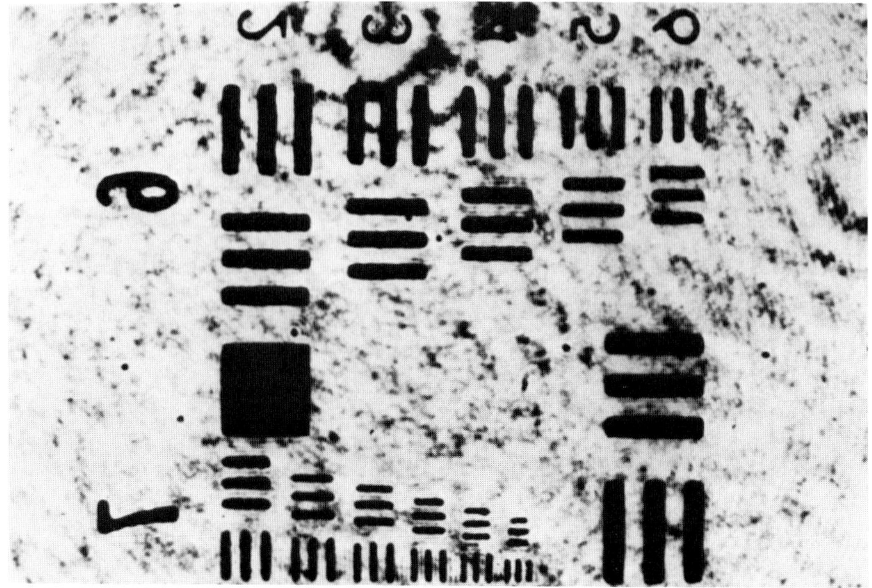

Figure 8.3. Holographic reconstruction of test target (×600).

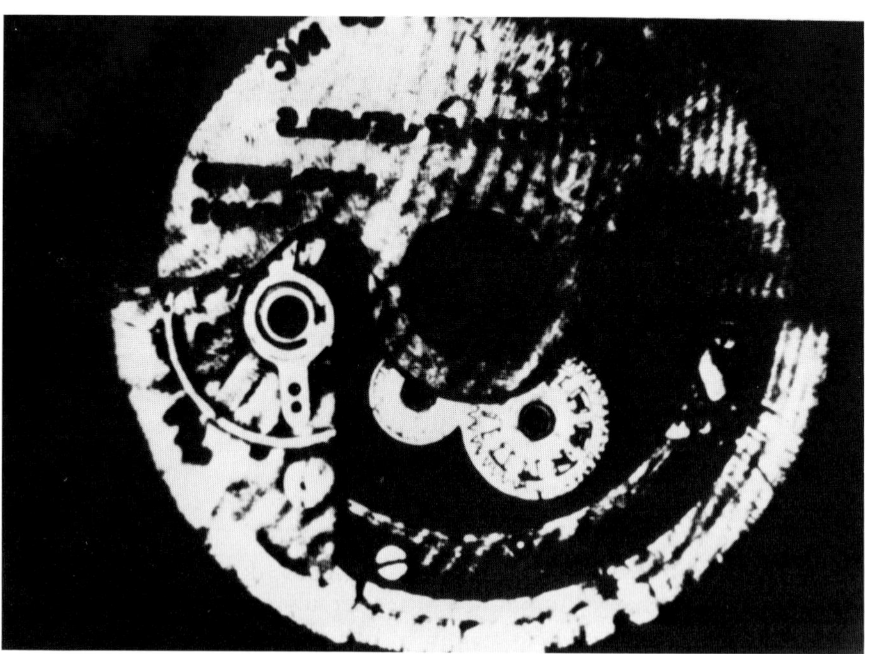

Figure 8.4. Full-field holographic reconstruction of a watch movement.

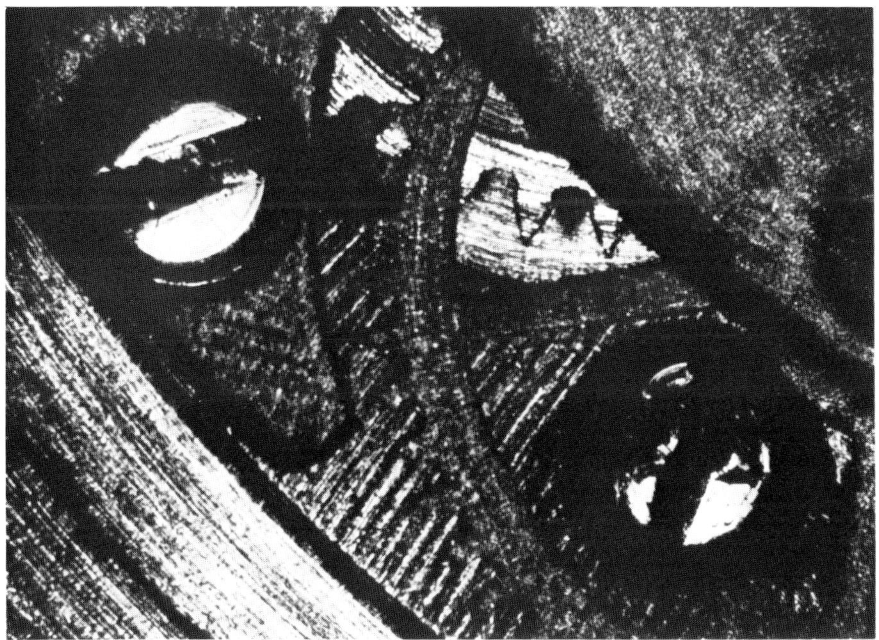

Figure 8.5. Enlarged view of center gears section of holographic image shown in Figure 8.4.

The calculated resolution of the system (classic Rayleigh resolution limit) is 4.0 μm. The photograph of the holographic reconstruction of the standard Air Force resolution target in Figure 8.3 shows the experimental results. The smallest bars in this target are on 4.4-μm centers. The resolution observed is clearly better than 4 μm. A more pictorial illustration of the performance of the system is shown in Figure 8.4 by the photograph of the watch movement. An enlarged view of the watchworks is shown in Figure 8.5. These photos demonstrate the capability of the system to record high-resolution images of semidiffuse objects. Thus, holographic documentation using such a system could have application to quality assurance, microbiology, documentation of jewelry and other valuables, and other fields where photomicroscopy is used.

Holographic inspection

With the successful application of the holographic documentation system to pretest characterization of high-energy laser test samples, it became clear that this system should have other valuable applications. A fortuitous meeting with a Silicon Valley entrepreneur led us in the direction of integrated circuit inspection. We learned from our interaction with workers in the integrated circuit field that a need existed for faster, less expensive means of detecting defects in integrated circuit photomasks and wafers. Instruments then available for this purpose were based on serial optical scanning and comparison of adjacent dice on a pixel-to-pixel basis. If the signal on the two optical pickups differed, one of the dice was assumed to be defective at this pixel location. These instruments were effective but costly and time consuming; full inspection of a photomask could take several hours.

Our objective was to apply the holographic documentation technology to obviate these drawbacks. We believed that a significant advantage would be realized by filtering out the perfect pattern information in the mask or wafer images, leaving only the defects to inspect, that is, defect enhancement. Conventional image analysis methods could then be used to count and characterize the defects. The entire process would take much less time, as only defect areas in the image would be addressed.

Some form of optical processing seemed an obvious approach and several methods were considered, some studied in detail. Included were image subtraction, matched filtering, and spatial filtering. Our theoretical and experimental studies led us to conclude that spatial filtering in the Fourier transform plane of a lens held the most promise for a practical defect enhancement system. The concept is illustrated in Figure 8.6. The first lens performs the Fourier transformation of the input image (an integrated circuit photomask or wafer). The distribution of light intensity in the Fourier transform plane is related to the spatial frequency content of the input image. Low spatial frequencies appear near the axis and higher frequencies lie farther out in the transform plane. An integrated circuit photomask diffracts light in a manner similar to that of a grating and the intensity pattern in the Fourier transform plane is a two-dimensional array of bright points. Low spatial frequency information, such as line width and line spacing, is located near the axis; high spatial frequency information, such as edge detail and defect shape, lies farther away from the axis. Defect light is usually spread out over large areas of the Fourier transform plane and is of low intensity. A filter designed to block the bright points suppresses the regular pattern information and allows this defect light to pass through. The second lens in Figure 8.6 performs the inverse Fourier

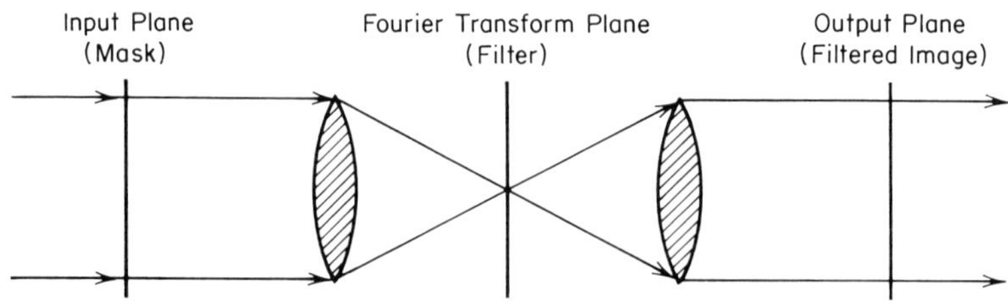

Figure 8.6. Optical processing by spatial filtering in the Fourier transform plane.

transform and reimages the input object onto the output plane. With an appropriate blocking filter in place a defect-enhanced image results. This spatial filtering technique is an example of parallel optical processing that has a significant speed advantage over serial scanning methods.

Practical implementation of this spatial filtering technique would seem straightforward. Numerous previous investigators had in fact studied spatial filtering for mask inspection and similar applications. These previous efforts had realized only limited success, however, due principally to difficulties arising from aberrations in the Fourier transform lenses. To obtain a nearly diffraction-limited image at the output plane of a Fourier optical system the lenses must be nearly aberration-free, and the lenses must be matched so that what aberrations there are will effectively cancel out. An acceptably low level of aberration is difficult to achieve with conventional glass optics, particularly with large apertures and low f/numbers. Five-inch-diameter Fourier transform optics required for integrated circuit work would be extremely expensive even if adequate optical quality could be achieved. Another problem has been the difficulty in producing effective blocking filters. Blocking filters may be computer calculated and generated mechanically, but adequately matching the filter to the aberrated Fourier transformed spectrum is difficult. Photographic generation of blocking filters has problems due to the limited dynamic range of typical photographic recording media.

Holographic optical processing

Solutions to the past problems encountered in applying Fourier transformed spatial filtering to inspection and defect detection were realized with the novel adaption of the holographic documentation system developed for the Air Force. The central observation was made that Fourier transform filtering and inverse transform imaging could be accomplished with a single, simple lens using the phase conjugation or reverse ray-tracing property of the hologram. The essential optical properties of the documentation system illustrated in Figure 8.2 as applied to Fourier transform spatial filtering are illustrated schematically in Figure 8.7. The test target, in this case an integrated circuit photomask, is illuminated by the subject beam in transmission. Polarization of the beam is rotated 90 degrees by the half-wave plate and reflected onto the lens by the polarizing beam splitting cube. The lens generates the spatial frequency spectrum of the mask at the Fourier transform plane and images the mask onto the hologram plate. When the processed hologram is illuminated by the conjugate reference beam the reversed rays recreate the Fourier transform spectrum and are reimaged by the lens in the mask plane. Aberrations in the lens are exactly compensated by reverse ray-tracing and a diffraction-limited image results. Because of this aberration compensation, an extremely high-quality lens is

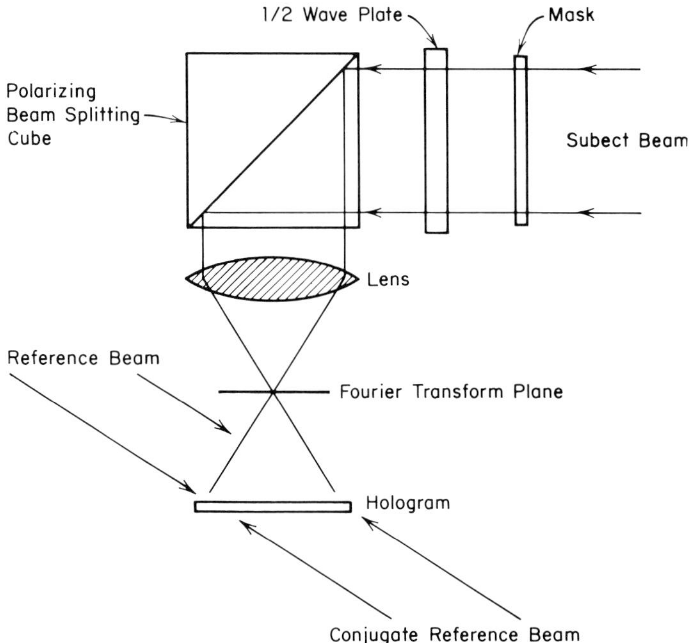

Figure 8.7. Schematic diagram detail of the $f/3.2$ holographic breadboard system as it is used for mask inspection and spatial filter production.

not required and a large field, low f/number system can readily be implemented. Processing of the image in the Fourier transform plane is now relatively straightforward.

The original holographic documentation system was designed to examine reflecting objects. A modification to inspect transmissive optics is shown in Figure 8.8. Two views of the system are shown in Figures 8.9 and 8.10. A second simplified $f/1.5$ breadboard system designed to examine only transmitting objects (masks and reticles) with higher resolution is illustrated in Figure 8.11. Photographs of two views of this system are shown in Figures 8.12 and 8.13. The holographic reconstruction of a bar test chart recorded with this system is shown in Figure 8.14. The classic Rayleigh resolution limit of the $f/1.5$ system for $\lambda = 514.5$ nm is approximately 1.25 μm and the detection limit is 0.6 μm. The 1-μm-wide bars are clearly resolved indicating that the system performed beyond the classic limit. Most of the experimental studies of defect detection were performed using the $f/3.2$ system and all results that follow are for this system.

Spatial filter generation

Photographs of two production integrated circuit photomasks are shown in Figures 8.15 and 8.16 and their Fourier transform plane intensity patterns

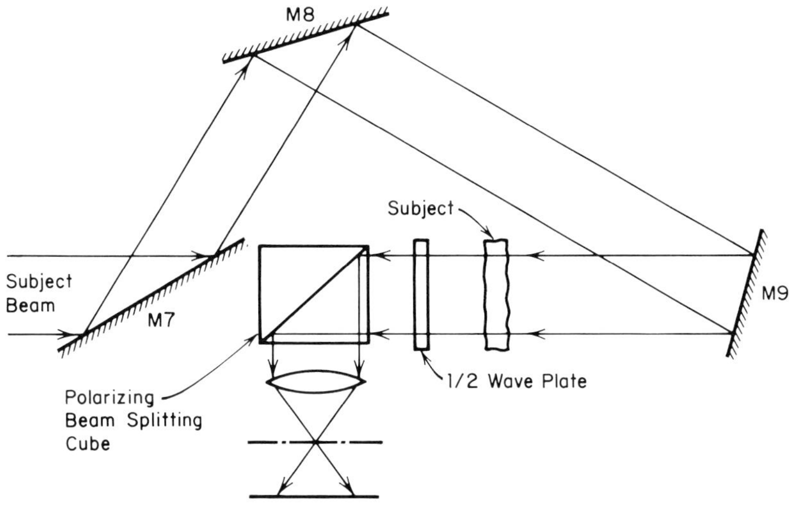

Figure 8.8. Modification of the $f/3.2$ holographic breadboard system for transmitting subjects (e.g., masks or reticles).

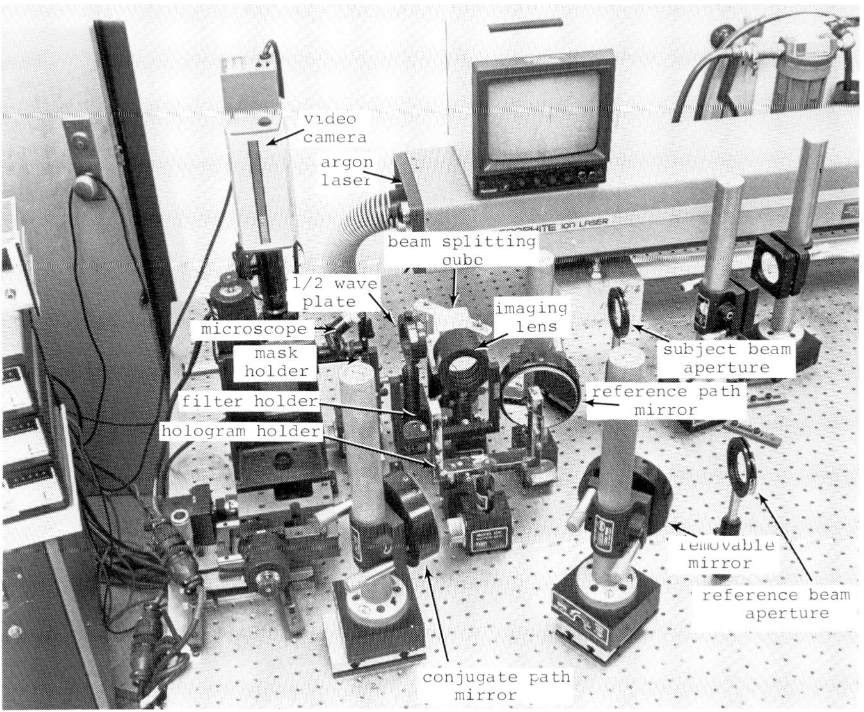

Figure 8.9. Photograph of the $f/3.2$ holographic breadboard system.

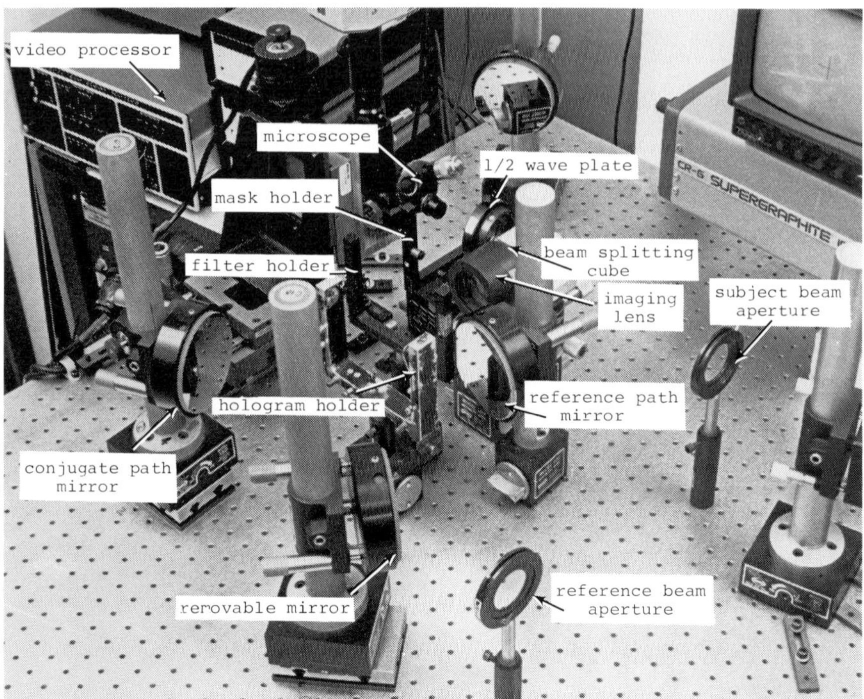

Figure 8.10. Detailed photograph of the $f/3.2$ holographic breadboard system.

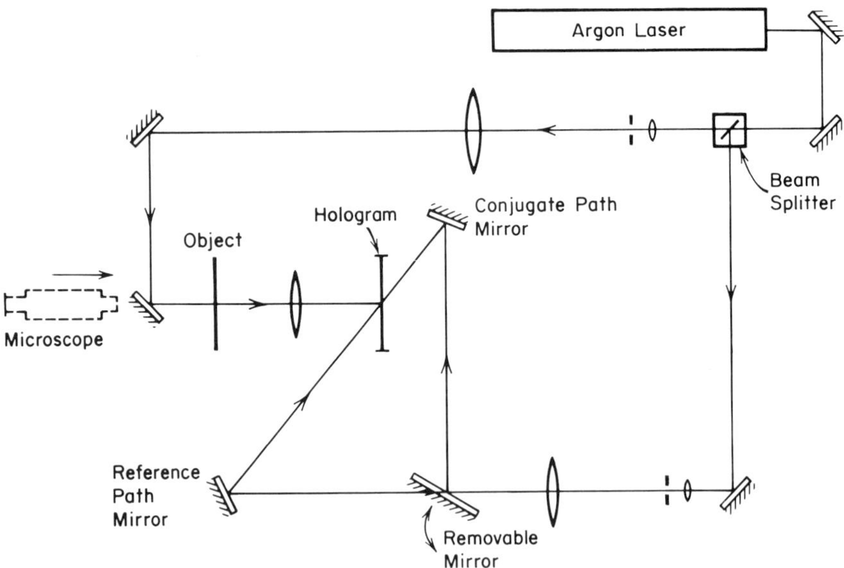

Figure 8.11. Schematic diagram of the $f/1.5$ holographic breadboard system for transmitting subjects (e.g., masks or reticles).

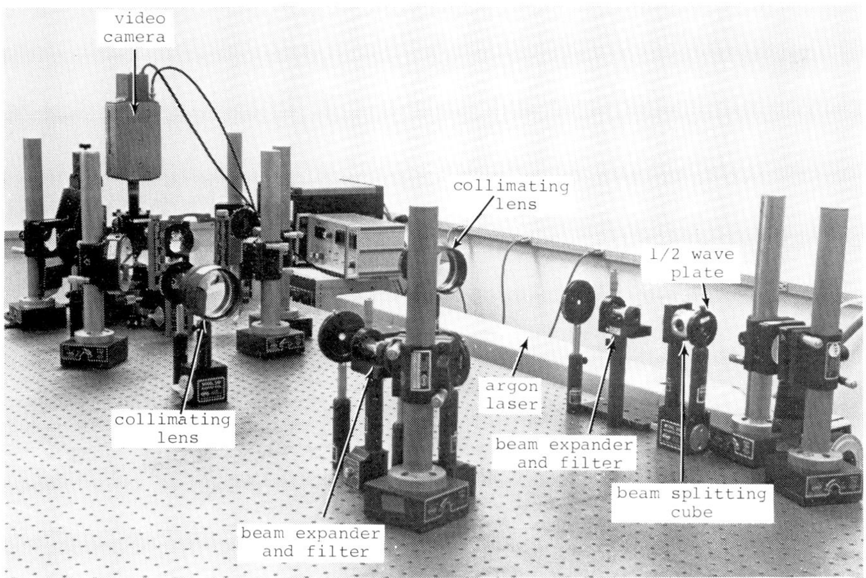

Figure 8.12. Photograph of the $f/1.5$ holographic breadboard system.

Figure 8.13. Detailed photograph of the $f/1.5$ holographic breadboard system.

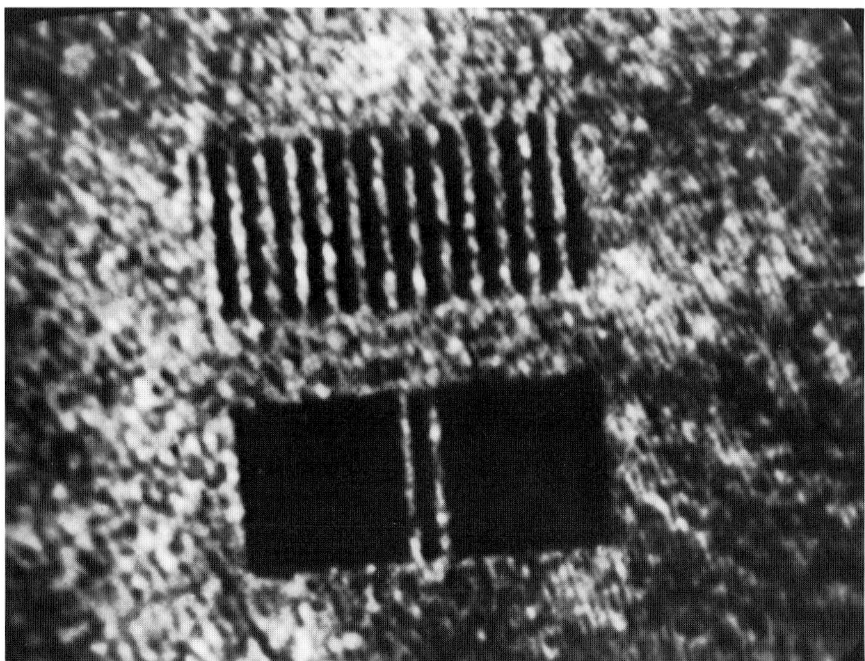

Figure 8.14. Photomicrograph of a holographic reconstruction on the $f/1.5$ bread-board system of a resolution bar test chart; the bars are $1\,\mu$m wide with $1\,\mu$m spacing.

are shown in Figures 8.17 and 8.18. Although the physical patterns of these two integrated circuit masks are visibly different, the Fourier transform intensity distributions are similar. The observation was made that the Fourier transform intensity patterns for masks from a mask set for a particular integrated circuit device are similar, whereas the intensity patterns for photomasks used to fabricate different integrated circuit devices are quite different. The Master Images Verimask™, shown in Figure 8.19, is used to evaluate the performance of mask inspection equipment. This mask has several different types of calibrated defects such as breaks in circuit lines, bridges between circuit elements, pinholes, pin dots, intrusions, and extensions. Defects range in size from 0.7 to 3.74 μm. The Fourier transform plane intensity pattern of the Verimask is shown in Figure 8.20. Note that this pattern is quite different from the intensity patterns for the other two masks.

Previous studies have shown that making an effective filter for a Fourier transform spatial filtering system is challenging. To be effective, the filter must be highly attenuating at the bright points and transmissive elsewhere. If the filter is not sufficiently transmissive in the dark areas, defect light will be attenuated also. An obvious method of making a blocking filter is to

Figure 8.15. Magnified view of a portion of a production IC mask (No. A). Each dice is approximately 3 mm^2.

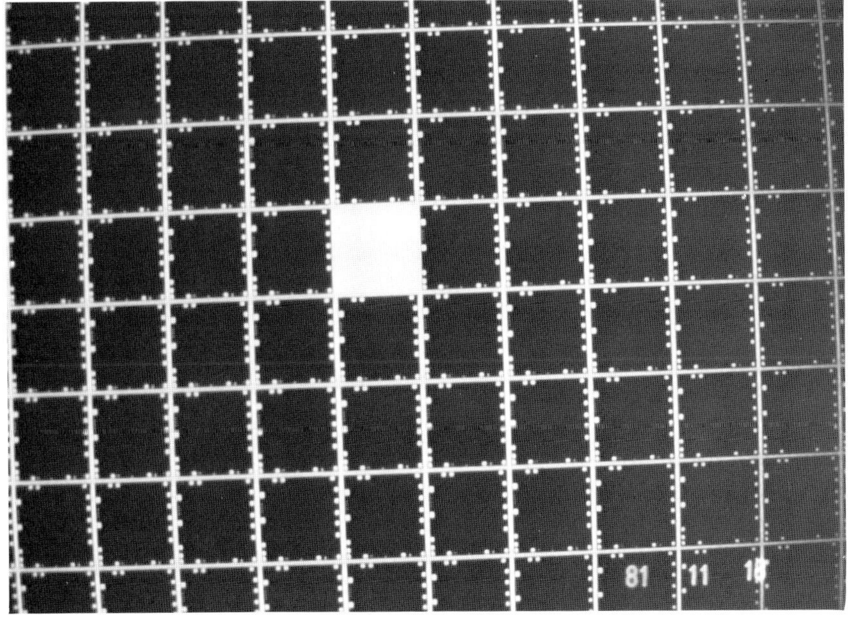

Figure 8.16. Magnified view of a portion of another production IC mask (No. B).

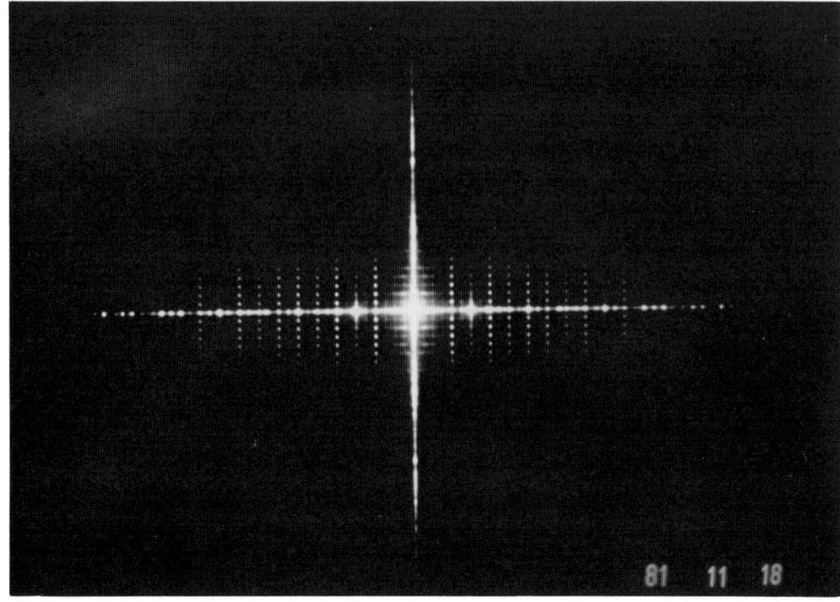

Figure 8.17. Optical Fourier transform of mask No. A.

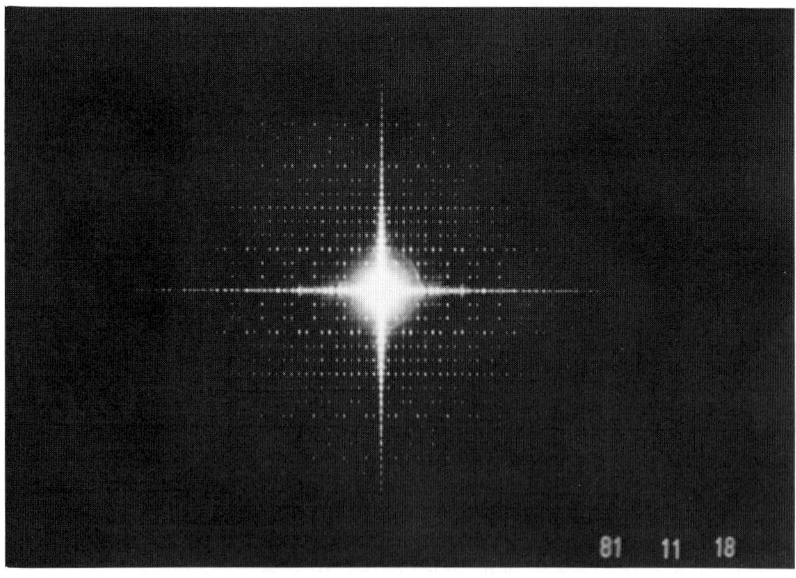

Figure 8.18. Optical Fourier transform of mask No. B.

Figure 8.19. Magnified view of a portion of the Master Images Verimask™.

Figure 8.20. Optical Fourier transform of the Verimask™.

expose a photographic plate in the Fourier transform plane. Negative development of this plate provides regions of high attenuation at the bright points and vice versa. The problem with this simple approach is the wide dynamic range of intensities in the Fourier transform plane that can be as much as 10 orders of magnitude. If the high-intensity central region of the pattern is properly exposed, the low-intensity high-frequency regions will be underexposed and the high-frequency components will not be effectively filtered. Conversely, if the plate is exposed long enough to record the high-frequency components, the central low-frequency region will be grossly overexposed and the defect light will be filtered excessively.

Again, the holographic optical processing method can be used to solve the filter generation problem. The dynamic range of the intensity pattern can be effectively compressed through a multistep filter generation process. The process is as follows: with a mask in place, a holographic plate is exposed to the Fourier transform pattern with exposure parameters set to record accurately the intensity distribution in the low-frequency region near the optical axis. After processing, this plate (stage 1 filter) is replaced in the Fourier transform plane and a hologram of the mask is made through this filter. The hologram is illuminated by the conjugate reference beam and a second filter plate (stage 2 filter) is exposed to the resulting Fourier transform intensity pattern. This pattern now has its low-frequency components attenuated due to the action of the stage 1 filter. The second stage filter can now be used to record a third stage filter and the process may be repeated as many times as necessary to produce a filter with the desired attenuation properties.

Once the desired filter is produced, a defect-enhanced image of the test piece is generated with this filter and the original unfiltered hologram is illuminated by the conjugate reference beam. Image rays are reversed back through the Fourier transform plane where they are filtered and through the lens where the inverse Fourier transform is performed. The filtered image is observed in the plane of the input test piece.

Note that with this process it is not necessary to make the filter with a defect-free reference mask. The test piece itself may be used as the defect light has insufficient intensity to expose the filter plate in the multistep process. This feature relies on the discrete nature of the Fourier transform intensity pattern generated by a test piece with the highly regular, highly repetitive pattern typical of photomasks.

First, second, and third stage dynamic range compressed blocking filters produced by this method for the Verimask are shown in Figures 8.21, 8.22, and 8.23.

Mask inspection results

The holographic reconstruction of a 2.01-μm pinhole defect in the Verimask is shown in Figure 8.24 and the second stage filtered image of the same

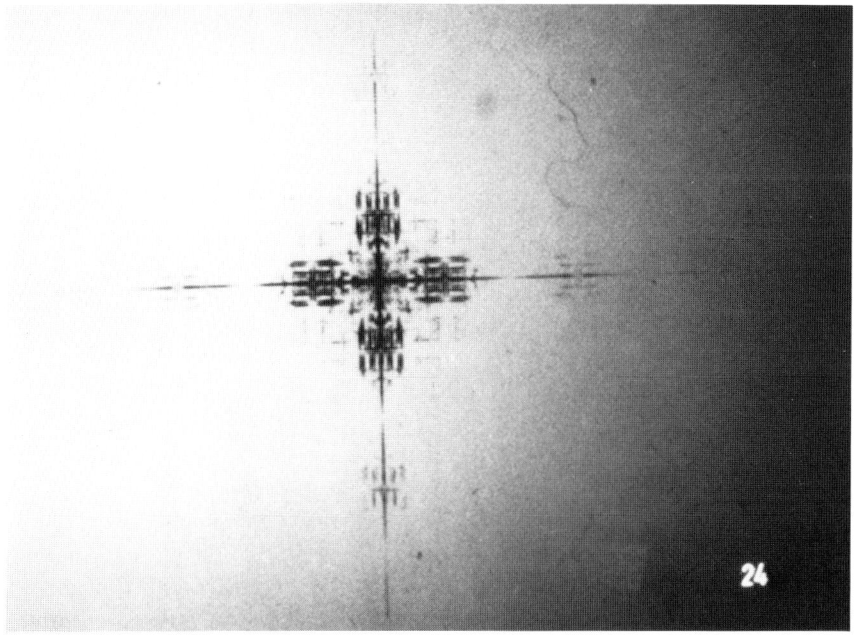

Figure 8.21. First-stage or initial blocking filter for the Verimask™.

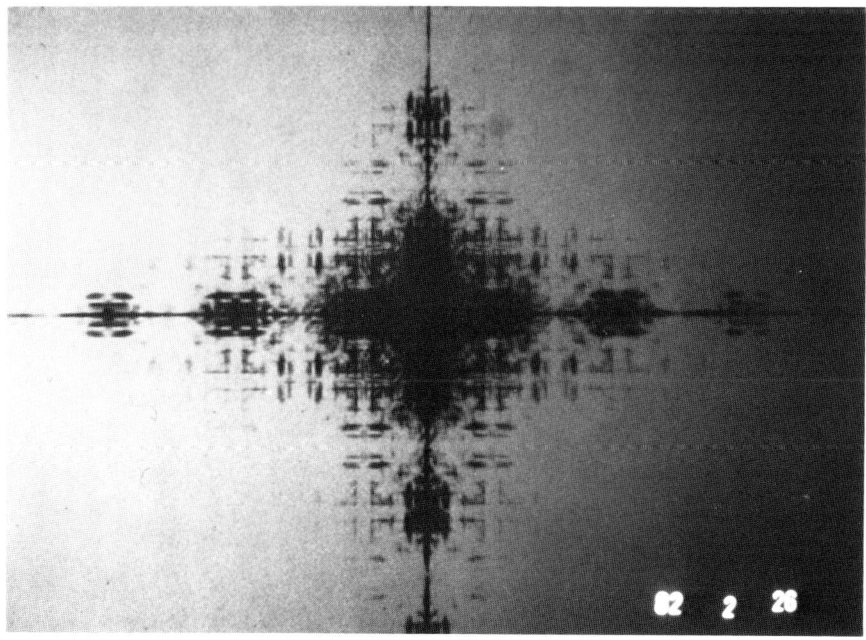

Figure 8.22. Second-stage dynamic-range compressed filter for the Verimask™.

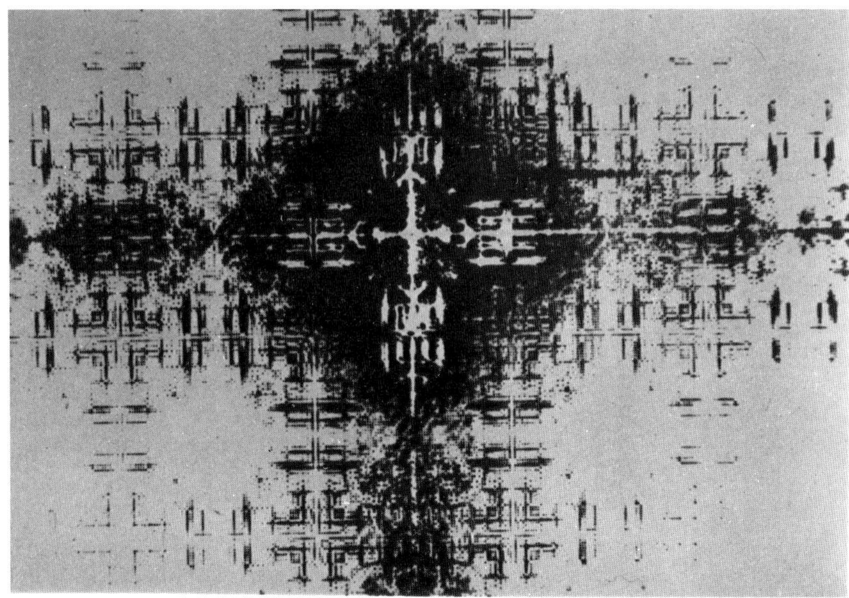

Figure 8.23. Third-stage dynamic-range compressed filter for the Verimask™.

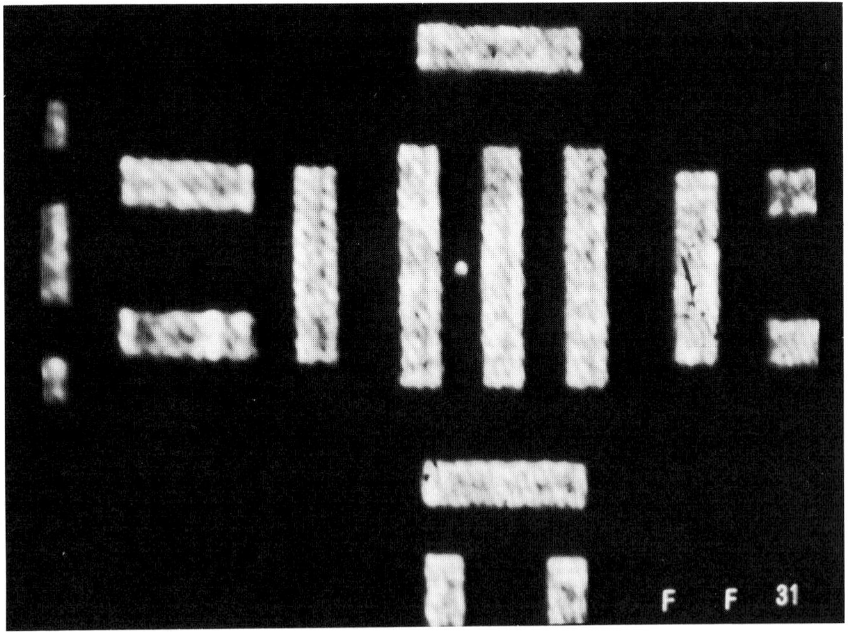

Figure 8.24. Photomicrograph of a holographic reconstruction of a 2.01-μm pinhole on the Verimask™.

defect is shown in Figure 8.25. Figures 8.26 and 8.27 show the unfiltered and filtered images of a 1.97-μm wide break. The filtered images clearly show the enhancement of the mask defects. In addition, dimensional variations in the mask pattern from dye to dye are also highlighted. This is illustrated by the bright ends and corners of the bars. The intensity enhancement of the defects is usually much larger than for dimensional variations; thus, the latter can usually be removed by simple video processing techniques such as increasing the contrast and reducing the brightness.

Figure 8.28 is a low magnification video monitor photomicrograph of a holographic reconstruction of the Verimask showing simultaneously four calibrated defects ranging in size from 1.95 to 2.50 μm. Significant enhancement of the defects was obtained with the use of the second stage filter. The enhancement is illustrated more clearly by removal of the background detail with a simple logarithmic amplifier in conjunction with the video monitor as shown in Figure 8.29.

Figure 8.30 is a photograph of a photomask consisting principally of parallel 100-μm wide bars. A video monitor photomicrograph of a section of this mask is shown in Figure 8.31. The region of the mask shown in Figure 8.31 contains subtle scratch defects that are extremely difficult to see

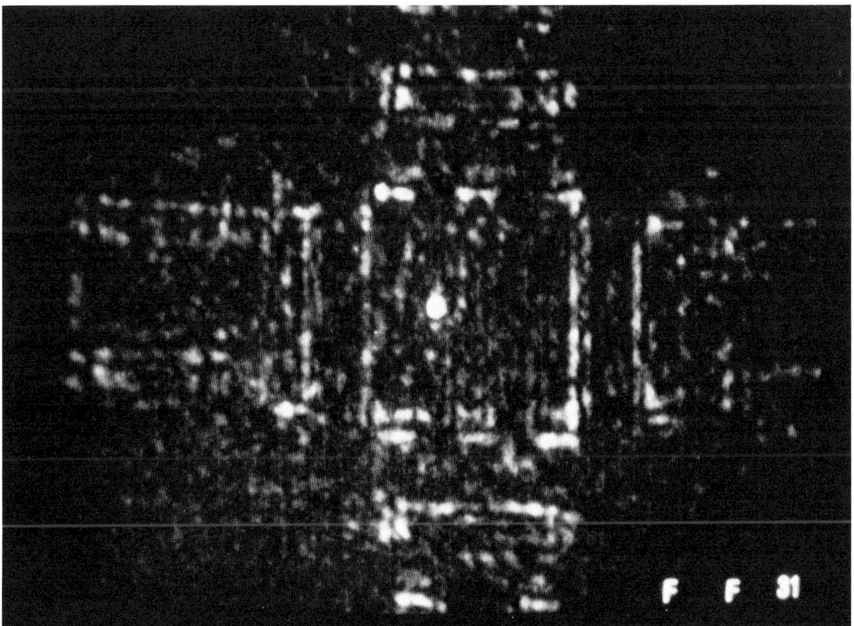

Figure 8.25. Same as Figure 8.24 except that a second-stage dynamic-range compressed filter (see Figure 8.22) was in the FT plane when the photomicrograph of the holographic reconstruction was made.

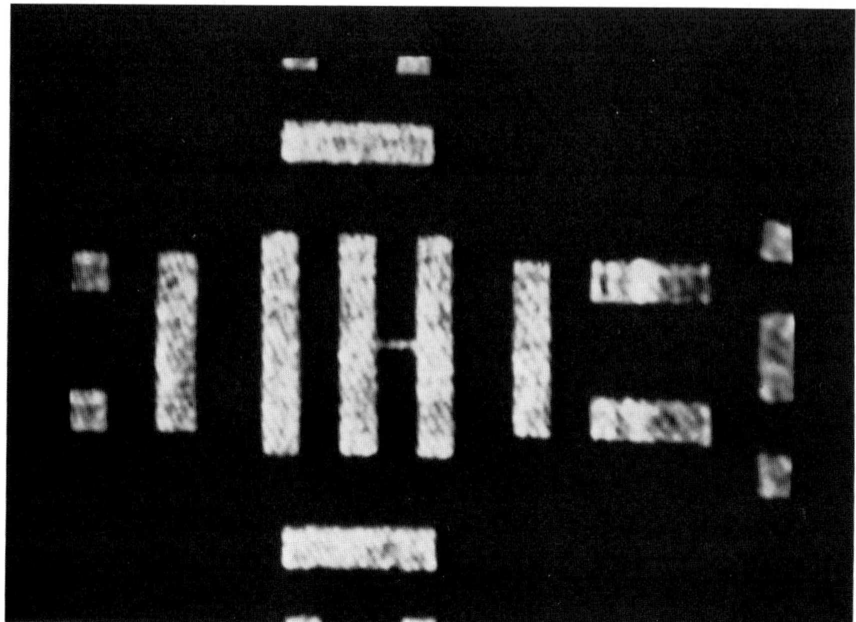

Figure 8.26. Photomicrograph of a holographic reconstruction of a 1.97-μm break on the Verimask™.

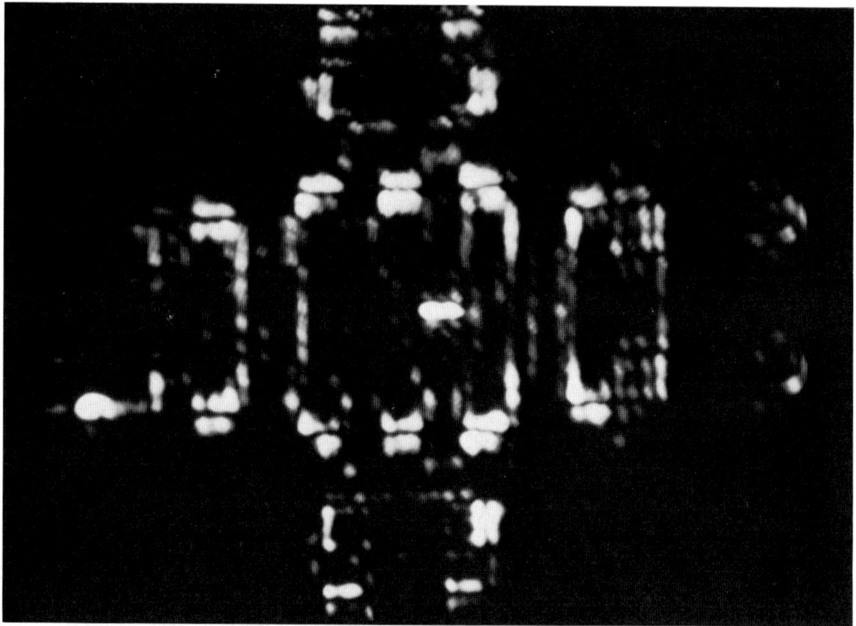

Figure 8.27. Same as Figure 8.26 except that a second-stage dynamic-range compressed filter (see Figure 8.22) was in the FT plane when the photomicrograph of the holographic reconstruction was made.

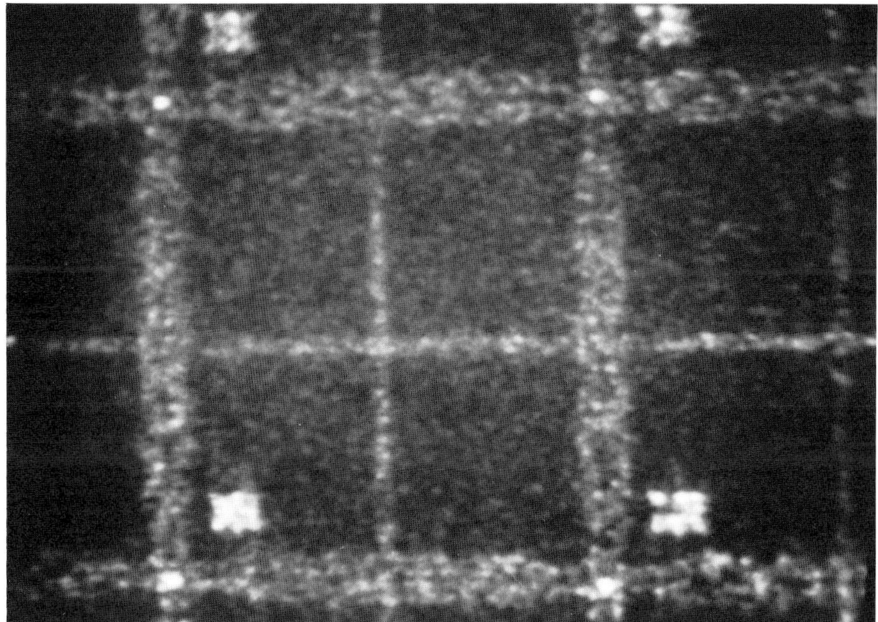

Figure 8.28. Low-magnification video monitor photomicrograph of a holographic reconstruction of the Verimask containing four pinhole defects ranging in size from 1.95 to 2.50 μm; a second-stage dynamic-range compressed filter was in the FT plane.

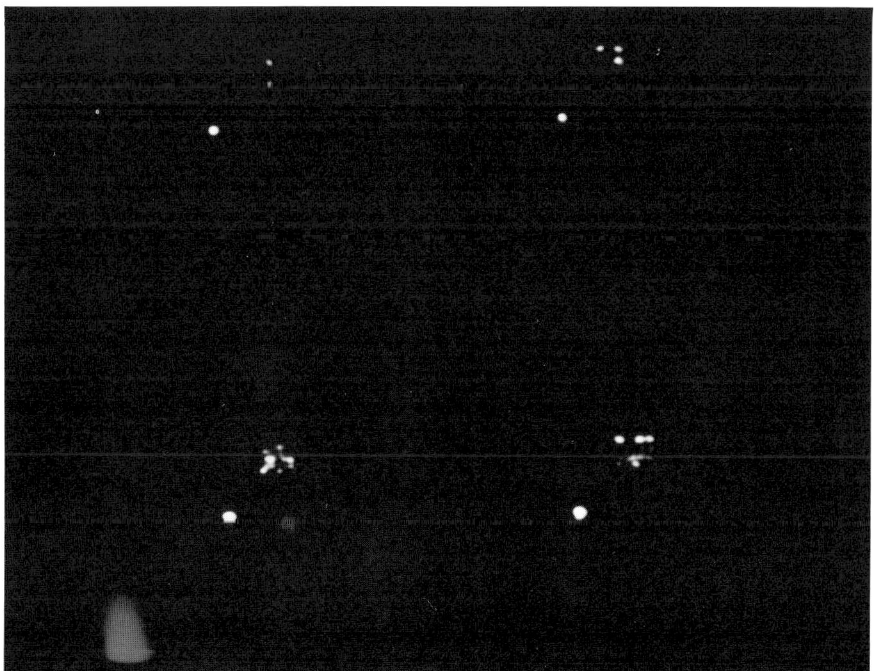

Figure 8.29. Same as Figure 8.28 except that a simple logarithmic amplifier was used in conjunction with the video monitor to nearly eliminate optical background noise.

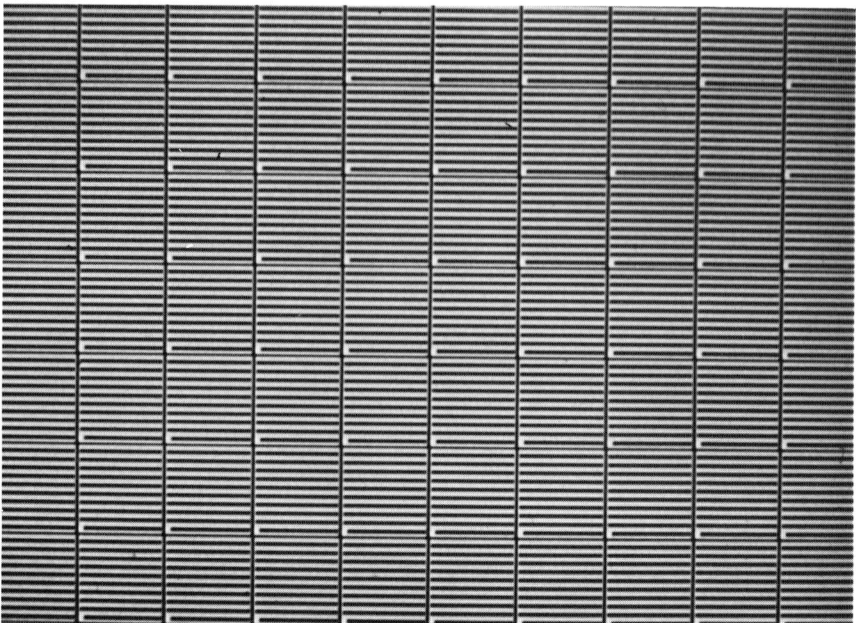

Figure 8.30. Photo of a mask containing 100-μm wide bars and subtle scratch defects.

Figure 8.31. Video monitor photomicrograph of area of mask shown in Figure 8.30 that contains scratch defects.

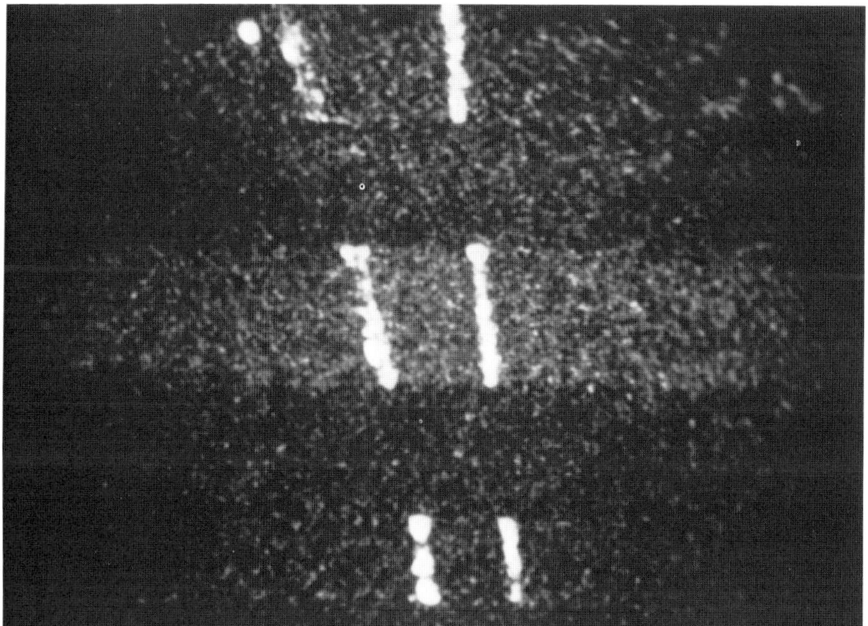

Figure 8.32. Same as Figure 8.31 except that a third-stage dynamic-range compressed filter was in the FT plane when the video monitor photomicrograph of the holographic reconstruction was made.

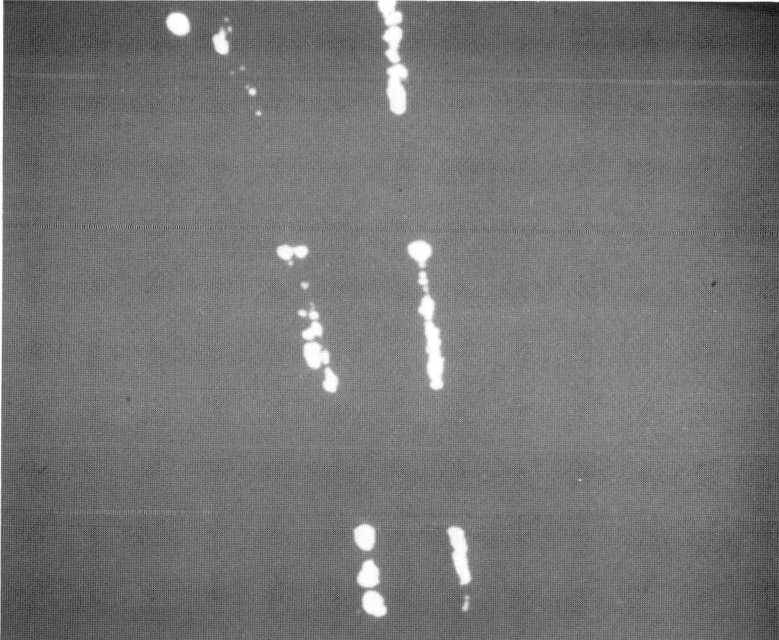

Figure 8.33. Same as Figure 8.32 except that a simple logarithmic amplifier was used in conjunction with the video monitor to nearly eliminate optical background noise.

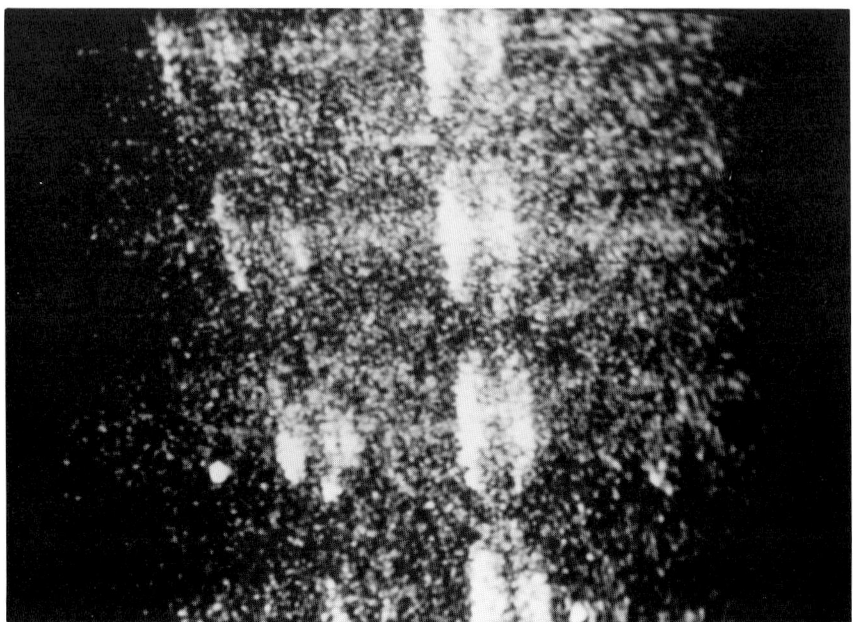

Figure 8.34. Same as Figure 8.32 except that the microscope was focused on the back surface of the mask substrate where a 5-μm dust particle was present.

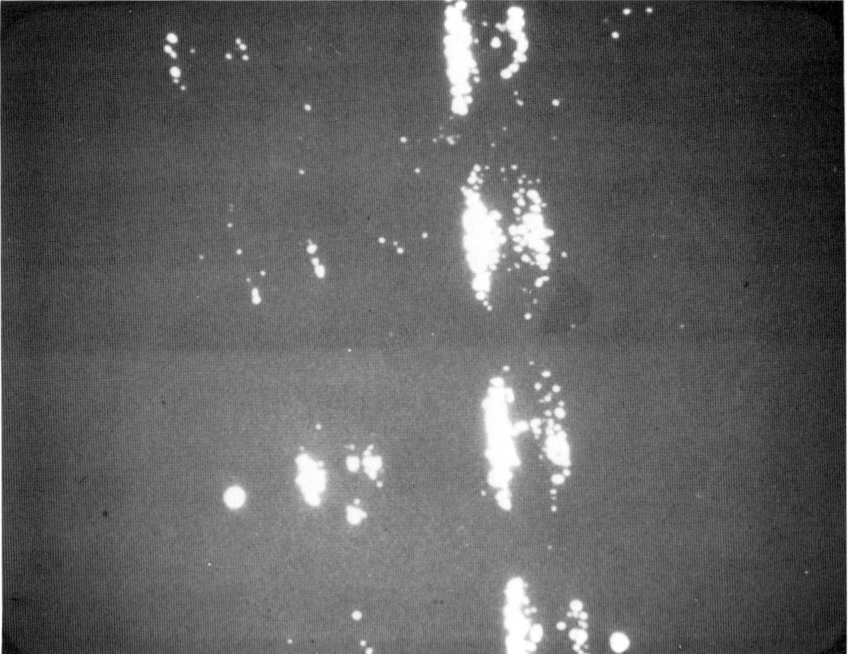

Figure 8.35. Same as Figure 8.34 except that a simple logarithmic amplifier was used in conjunction with the video monitor to nearly eliminate optical background noise.

under normal optical microscopy. These defects become readily apparent with the use of holographic defect enhancement as illustrated in Figure 8.32. The defects are made more visible with the use of simple video processing to remove the background noise as illustrated in Figure 8.33. An important feature of the holographic system is the capability of recording depth information in the image. This is illustrated by focusing the microscope on the back surface of the 100-μm-wide bar mask (Fig. 8.34). The scratches are observed out of focus, but we now see a 5-μm dust particle on the back surface. The dust particle is more apparent in Figure 8.35 with the background noise removed. The depth capability of the holographic inspection system makes possible the inspection of pellicle-coated masks or unobstructed parts of sublayers in integrated circuit wafers.

Other concepts

Our experimental effort concentrated on Fourier transform plane spatial filtering. The necessity to record these filters is an obvious drawback of the technique. One promising method of eliminating the need for the filters was to record the hologram in such a way that only defect light was efficiently reconstructed. We believed this could be done by placing the hologram in the Fourier transform plane and reducing the intensity of the reference beam to the level of the defect light. The diffraction efficiency of the hologram should thus be high for the defects and low for the regular pattern information, which is of much higher intensity and much larger than the reference beam intensity. This method was considered but not studied due to time and funding constraints. Another disadvantage of the technique described here is the need to record a hologram of each article to be inspected. We also considered the possibility of using real-time holographic recording material such as a photorefractive crystal to eliminate this requirement. Both of these concepts have recently been demonstrated by researchers at Stanford University (6).

Conclusions

A holographic documentation system was developed and used successfully to document the surface microstructure of high-energy laser optical component test samples. Recording a single high-resolution holographic image of the sample was much more efficient than the use of conventional photomicroscopy methods. This system was adapted to the inspection of integrated circuit photomasks and wafers with the use of novel holographic optical processing concepts.

Acknowledgments

The work described in this paper was performed principally by the following individuals: Richard L. Fusek (now with Insystems, Inc., San Jose, CA), Kevin Harding (now with The Industrial Technology Institute, Ann Arbor, MI), James S. Harris (now with JSH Optics, Centerville, OH), Steven C. Gustafson (UDRI), and John Murphy (UDRI). This work was supported by the U.S. Air Force Weapons Laboratory, Contract No. F29601-79-C-2007, B and B Associates, and Insystems, Inc.

References

1. Fusek, R. L., Harding, K., Harris, J. S., and Murphy, J. Holographic Documentation Camera for Component Study Evaluation. SPIE Proceedings, Meeting on High Power Lasers and Applications, Los Angeles, CA, February 11–13, pp. 186–195 (1981).
2. Fusek, R. L., Lin, L. H., Harding, K., and Gustafson, S. *Optical Engineering* **24**(5), 731–734 (1985).
3. Lin, L. H., Cavan, D. L., Howe, R. B., and Graves, R. E. *SPIE Proceedings* **538,** 110–116 (1985).
4. Lin, L. H., Billat, S. P., Cavan, D. L., Howe, R. B., Graves, R. E., and Fusek, R. L. *SPIE Proceedings* **775** (March 1987).
5. Billat, S. P. *SPIE Proceedings* **774** (March 1987).
6. Ochoa, E., Goodman, J. W., and Hesselink, L. *Optics Letters* **10**(9), 430–432 (Sept. 1985).

9
Application of holography to underwater visual inspection

JOHN WATSON

There is a growing number of industrial applications of holography where the required end product is a high-resolution hologram of a particular scene of interest. From this hologram dimensional analysis can be performed directly on an image reconstructed in real space (1). This field of holography has now come to be known as *hologrammetry* and is particularly useful when the inspection site is in a hazardous environment or in an environment where access is difficult. Current applications include visual inspection of nuclear fuel elements (2–4) or sections of underwater structures (5–9) and bubble chamber holography (10). The work could easily be extended to cover areas such as marine growth, underwater bubble fields, and constructional archiving.

Our interest, at Aberdeen University, has been with the application of hologrammetry to high resolution measurement of subsea components and structures. This work involves the recording of holograms of objects submerged in water with subsequent replay of the real image in air. Inevitably this process results in an image that will have a resolving power below that of the equivalent situation in air. The image will also possess optical aberrations brought about by recording in one refractive index medium and replaying it in another. In this chapter we report on some of the work carried out in investigating the nature of these problems.

Concept of hologrammetry

The basis of hologrammetry as a means of high-resolution visual inspection is the holographic recording of the scene of interest with the subsequent replay of the processed hologram in the real image mode of reconstruction, (1, 11, 12). Reconstruction of the real image produces an image that is reversed left-to-right and back-to-front when viewed from the space in front of the hologram. Such a representation of the image is known as *peudoscopic*. To allow precise dimensional measurements to be made the real image should be aberration-free and possess high optical resolution.

In hologrammetry, there are several recognized factors that can give rise to optical aberrations and, ultimately, degrade image fidelity. Such factors include:

1. Nonuniform distortion of the fringe pattern recorded in the emulsion as a result of chemical processing
2. The optical quality of the emulsion substrate
3. Variations between recording and reconstruction geometries
4. Variations between recording and replay wavelength
5. The quality of the reconstruction beam

These factors affect the ability to produce optimum conditions for wavelength reconstruction but can, with reasonable precautions, be controlled to a degree sufficient to produce high-resolution images (2–4).

The requirement of recording underwater with subsequent laboratory reconstruction in air, however, introduces additional factors upon which image fidelity may depend. Such factors include:

1. Thermal gradients in the water
2. Turbulence in the water
3. Salinity gradients
4. Polarization effects
5. Scattering
6. Absorption
7. Mismatch between the refractive index of the medium in which the hologram is recorded and that in which it is replayed

Many of these effects have been discussed in previous chapters (Chapters 5 to 9). In this chapter we are most concerned with the potential resolving power of the underwater hologram and the influence of the refractive index mismatch.

Resolution of underwater holograms

To compare and contrast the resolving power of holograms recorded of underwater objects with the equivalent holograms taken in air the optical arrangement shown in Figure 9.1 was used, both with and without the observation tank in place. The target for all resolution measurements was the standard USAF (1951) resolution chart possessing a series of vertical and horizontal bars in the range 1 to 228 lp/mm. The resolution of the reconstructed real image was measured using a traveling microscope fitted with a ×10 Ramsden eyepiece with an overall system magnification of ×20. All holograms were recorded on Agfa holographic plates type 8E56HD and processed using a pyrogallol-based developer (Agfa formulation GP 62) and bleached in a parabenzoquinone (PBQ)-based bleach (Agfa formulation GP 432).

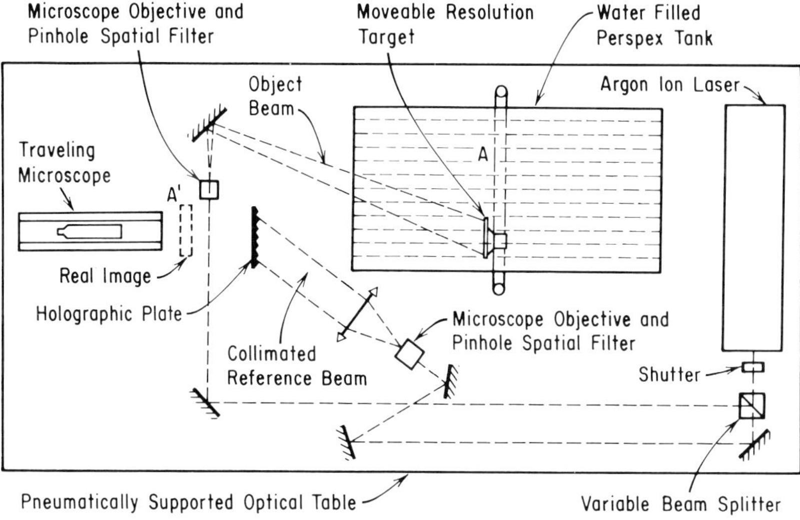

Figure 9.1. Optical arrangement used to compare and contrast the resolving power of holograms recorded at underwater objects with the equivalent holograms taken in air.

Holograms were recorded with the unexposed plate in the plate holder and the resolution chart in position A in the optical arrangement. After processing according to the method outlined previously the hologram was rotated through 180 degrees and illuminated through the back of the plate. The corresponding real image is reconstructed in position A'. Collimation of reference and reconstruction beams was accomplished using a lens with a focal length of 300 mm and aperture of $f/5$. The wavefront accuracy is $\lambda/8$ over the central 38 mm of its aperture at 514 nm. To illuminate as large an area of the holographic plate as possible the entire lens aperture was used to expose a roughly elliptical area of 65×75 mm of the film. The lens was collimated to an estimated divergence of no more than 2 mrad. A reference beam to optic axis angle of 30 degrees kept the spatial frequency of the system below that of the cut-off frequency of the film.

The laser used in the experiments was an argon ion (Lexel type 90-4) delivering up to 1.5 W single frequency at 514 nm. The entire set-up was mounted on a vibration isolated table.

A number of holograms were recorded to monitor the optical resolution achievable when recording the holograms underwater. Initially, two holograms were taken under the conditions outlined, but with no observation tank in position. Therefore, the holograms were recorded entirely in air to establish a reference point. These holograms were taken at target-to-film distances of 550 and 1000 mm, respectively. A second pair of holograms were recorded at the same target-to-film distances but with a perspex observation tank in place. In this case the perspex wall was nominally 10-mm thick and the distance between the front wall of the tank and hologram plane was 240 mm. A third pair of holograms were recorded as previously but with turbid water in the observation tank. In this situation the holograms were recorded at in-water paths of 300 and 750 mm, respectively. In all cases the film, perspex interface, and target were parallel to each and on the same optic axis.

The resolution of a holographic image is usually defined as the ability to distinguish two points on a hologram separated by distance r, given as,

$$r = 1.22\lambda z/D$$

where λ is the reconstruction wavelength, z is the separation between hologram and reconstructed image, and D is the effective aperture of the hologram. This equation is usually expressed as a resolving power (R) in line pairs per millimeter (lp/mm) by reciprocating and dividing by 10^3. In holography the presence of speckle effects introduced by the coherence of the light and the finite aperture of the viewing system influences the resolving power. The speckle size sets the lower limit to the resolution. In practice, R is increased by a factor of 2 to 3 to take account of this.

The experimentally obtained resolving powers are shown in Table 9.1. As a reference point a resolution of 57 lp/mm was measured directly on the original resolution chart using the measuring microscope described earlier when illuminated by reflected laser light at 514 nm.

The figures obtained for resolving power of underwater holograms show a decrease over the reference holograms recorded in air as, first, a perspex interface is placed in the optical path and then, second, a water interface is added to the path. The ultimate reduction in resolving power is from 22 to 18 lp/mm. It is not known at this stage whether this reduction is absolute or relative. These figures should be contrasted with those obtained by underwater photogrammetry, which indicate a resolving power of around 0.5 lp/mm for similar viewing conditions in sea water (13).

Table 9.1. Resolving power measured from underwater holograms

Location of target	Total film to target dist (mm)	Path distance of target in water (mm)	Measured resolution (lp/mm)
In air	550	—	22
In air/in tank[a]	550	—	20
In water/in tank[a]	550	300	18
In air	1000	—	9
In air/in tank[a]	1000	—	8
In water/in tank[a]	1000	750	7

[a] This distance includes a film plane to tank separation of 240 mm and a nominal tank wall thickness of 10 mm

Figure 9.2. Photograph (enlarged) taken from real image reconstruction of a hologram of a stress-induced crack (recorded in water, replayed in air).

In the field, measurement of resolving power is not a very meaningful figure considering that the object may be rough, poorly reflecting, and of low contrast. Being able to measure a particular set of bars on a resolution target does not really help in determing whether or not a particular surface feature can be visualized using holography. To illustrate this a hologram was taken of an engineering test piece: a polished titanium block with a stress-induced crack in it. The hologram was taken under the conditions outlined with a total object-to-film distance of 550 mm and a distance in water of 300 mm. A photograph taken from the reconstructed real image is shown in Figure 9.2. The crack, which was measured to be 40 μm across the root, is clearly visible.

Image shift in underwater holograms

When a hologram is recorded underwater and replayed in air, the reconstructed image suffers from optical aberrations due to the difference in refractive index between the two media. For points on-axis the image replays closer to the film plane in the simple ratio of the refractive indices. This latter fact will be modified by the presence of the perspex interface.

Table 9.2. Image shift measured from underwater holograms

Object position		Image position		Image shift	
On-axis (mm)	Off-axis (mm)	On-axis (mm)	Off-axis (mm)	On-axis (mm)	Off-axis (mm)
417	0	309	0	108	0
417	230	288	227	129	13
417	435	149	388	268	43[a]
		194	432	223	3[a]

[a] In this case two image positions were obtained, the first one corresponding to the vertical bars on the resolution target and the second one corresponding to the horizontal bars of the target.

Measurements of the image shift were made using the optical arrangement shown earlier, but this time provision was made to move the target position laterally with respect to the optic axis. For each hologram a ground glass screen was used to view the real image and the reconstructed image position determined. Table 9.2 shows the measured image shifts for a number of target locations.

The data show that for on axis points the image shift is in accordance with the shift predicted from a simple refractive index ratio. As the target is moved laterally with respect to the optic axis, however, the measured image shift increases beyond that expected by simple theory. The reasons for this being that now the light rays traveling from the hologram to the image position are traversing paths that are substantially different from those encountered in recording. In particular, the refraction of light encountered at the air/perspex/water interface during recording does not occur in replay. The replayed rays do not converge to the same point from which they emanated in recording. As a consequence of this, two distinct image points are formed on reconstruction: one for the horizontal resolution bars and one for the vertical resolution bars. Therefore, the image is astigmatic. As the object is moved further from the optic axis the difference between the two image positions increases.

Conclusions

Initial studies of the resolving power obtained from holograms recorded of objects submerged in water and replayed in air shows that good resolution can be obtained when the target and film are on the same optic axis. Although suffering some degradation over the equivalent situation in air the resolving power obtained from the underwater holograms leads to some optimism regarding future application of the technique. The resolving power obtained is superior to that obtained by stereo photogrammetry under similar conditions. Holograms, however, taken of off-axis targets would be expected to show a poorer resolving power than the on-axis case.

The key to the future application lies with the understanding of the nature of the optical aberrations introduced because of the refractive index change between recording and replay. The most significant of these aberrations would seem to be astigmatism, how to correct for it, and its influence on the resolving power of the system. It is on these areas that current work is concentrating.

References

1. Watson, J. *Optoelectronics,* Van Nostrand Reinhold UK, (1987).
2. Tozer, B. A., and Webster, J. M. *Proc Electro-Optics/Laser International '80 UK,* Brighton, UK, (1980).
3. Glanville, R., Lewin, L., Little, M. J., Tozer, B. A., and Webster, J. M. *Proc Electro-Optics/Laser International '84 UK,* Brighton, UK, (1984).
4. Tozer, B. A., Glanville, R., Gordon, A. L., Little, M. J., Webster, J. M., and Wright, D. G. *Proc SPIE* **523**: Applications of Holography, (1985).
5. Watson, J. *J Soc Underwater Tech,* **Winter**, 16–20, (1981).
6. Watson, J., and Britton, P. W. *Proc SUBTECH 83: Design and Operation of Underwater Vehicles,* London, UK, (1983).
7. Watson, J., and Britton, P. W. *Optics and Laser Tech,* 215–216 (1983).
8. Britton, P. W., and Watson, J. *Proc Electro-Optics/Laser International '84 UK,* Brighton, UK, (1984).
9. Watson, J., and Britton, P. W. *J. Photog Science,* **33,** 167–173 (1985).
10. Proc Meeting on Application of Holographic Techniques to Bubble Chamber Physics, Rutherford Appleton Labs, UK (1981).
11. Caulfield, H. J. *Handbook of Optical Holography,* Academic Press (1979).
12. Hariharan, P. *Optical Holography,* Cambridge Univ Press (1984).
13. Turner, J. D. *J Soc Underwater Tech,* **Winter**, 7–13 (1983).

IV
Holographic recording materials

10
Dry polymer for holographic recording

SERGIO CALIXTO

Optical storage can be performed by three methods: direct image, holographic storage, and digital storage. There is some overlap between the material requirements for each of these methods. For holographic storage the ideal material should have characteristics such as high sensitivity and high resolution, broadband response (400 to 700 nm), and grainless structure. The desirable development and fixing processes should be dry and preferably performed in a short time. The performance of the photosensitive medium should be repeatable and present a long shelf-life. Insensitivity to radiation and solvents after processing, availability in various thicknesses and sizes, eraseability, infinitely cyclable, and inexpensive are other good features.

Although the ideal holographic material should present the characteristics mentioned, few applications demand this perfect material. At present there exists a variety of materials to perform holographic storage (1). Among them are photographic plates, photochromic materials, thermoplastics, ferroelectric crystals, photoresists, and dichromated gelatin. Each one of them presents desirable or undesirable characteristics, but without doubt the most popular material is the photographic plate because it presents high sensitivity, resolution, familiarity, availability, and low cost. This last advantage is gained from batch fabrication procedures. The most noticeable drawback of the photographic plate is the wet development process that has to be done; it is messy and takes time.

In the present work are reported introductory results when dry acrylamide–polyvinyl alcohol films were used in holographic storage. The studies comprise the behavior of the transmittance and the diffraction efficiency as a function of exposure and the beam ratio. It was found that the material presents a pronounced reciprocity failure. Diffraction efficiencies of about 10 percent can be obtained when combined recording beams power is about $230\,\mu\mathrm{W/cm}^2$. As the polymeric mixture presents a self-developing process it can be read just after the exposure time or it can be fixed by illuminating it with an ultraviolet source. Applications in the field of Fourier holograms and multiple image by holography are shown.

Next, a brief introduction about photopolymers applied to holography follows. This is done to have a basis with which to compare the performance of the new polymeric mixture proposed in this work.

Photopolymers were among the first materials used to displace the photographic plate in holography. Close and colleagues (2) mentioned the use of a polymeric solution formed with acrylates and sensitized with methylene blue (MB); exposures needed were about 1 to 30 mJ/cm^2 at 694 nm with diffraction efficiencies of about 45 percent. The scattering noise, however, reduced the signal to noise ratio and the mixture presented a short shelf-life. This rapid decrease of photopolymerizing activity is due to the formation of sulfones in the presence of water (3). The fixing process was done by optical methods, that is, exposure to ultraviolet radiation that converts the dye to a colorless form. Later Jenny (4–6) studied different forms of this mixture, finding sensitivities of about 0.6 mJ/cm^2. The modulation produced by the writing light in the material was present in the volume or the surface, or both, depending on the spatial frequency of the interference pattern. Diffraction efficiencies of about 80 percent were achieved, but at spatial frequencies greater than 1500 lines/mm the signal to noise diminished considerably. To fix the holograms Jenny (4) describes two methods, one is to flash the photopolymer holograms with light from a xenon flashlamp (3 mJ/cm^2), and the other, named thermal fixing, involves keeping the hologram in the dark until the catalyst is completely deactivated. One drawback of this photopolymer is that some of the composing materials are poisonous and volatile. Also, the presence of surface gratings sometimes outweights the bulk effects. Another improvement on the mixture that uses acrylamides was presented by Van Renesse (7). He added methylene-*bis*-acrylamide to speed up the polymerization reaction. A molecular space lattice of acrylamide chains is formed and in this the methylene-*bis*-acrylamide builds cross-links to form a copolymer with a transparent and rigid characteristic. Sensitivity of this mixture is about 5 mJ/cm^2. Drawbacks with this material seem to be the lack of adhesion of the photosensitive material to the substrate and crystallization of the chemicals. The recorded hologram could be kept in good condition for only 1 week. An improvement of this mixture was cited by Sadlej (8). Later Sugawara and colleagues (3) reported a method that uses a mixture of acrylamide and methylene-*bis*-acrylamide with a photoreductant, such as acetylacetone or triethanolamine (TEA). Diffraction efficiencies of about 65 percent at exposures of 50 mJ/cm^2 were obtained. Photosensitivity was maintained for more than 80 days. The solution was enclosed between two plates with a thin film as spacing to obtain the desired thickness, as in the previous cited works. No optical fixing of the photopolymer was performed in this study. A different approach was taken by Jeudy and Robillard (9), who used a photochrome that changed its absorption band when excited by ultraviolet light having a wavelength of 0.3 to 0.4 μm. When this material is in its excited state it acts as a sensitizer for the photopolymerization process in the same way as ordinary dye sensitizers. In that case the sensitizing action will only exist when irradiation with ultraviolet light occurs and a fiixing process is not needed. Diffraction efficiencies of about 80 percent were obtained when an energy density of $100 \, \text{mJ/cm}^2$ was used. One drawback of this material is that it should be used within 4 to 5 days after preparation.

Following the procedure laid by Jenny (4), Sadlej and Smolinska (10) added to the photosensitive system a protective polymer. In this way they produced stable photosensitive layers that were used to make diffraction gratings and holograms. The protective polymers were polyvinylacetate, methylcellulose, polyvinyl alcohol, and gelatin. To control the rate of photoreaction of MB they used p-toluene–sulphinic acid sodium salt. Dry layers with a thickness between 15 and $50 \, \mu\text{m}$ were fabricated. For some mixtures they obtained volume effects, but unfortunately diffraction efficiencies of only 0.5 percent were obtained. In the case where a surface relief modulation was present diffraction efficiencies of about 4 percent were measured. Sensitivity was about $10 \, \text{mJ/cm}^2$.

Besides the acrylamide base photopolymers described, other types of photopolymers have been mentioned in the literature. One of them consists of a liquid acrylic monomer, a cellulosic binder, a photoinitiating system, and a plasticizer (11–16). This type of photopolymer is presented in a dry form and was made by E. I. Dupont de Nemours. Thick layers $(1-150 \, \mu\text{m})$ of this material can be produced. Poor spatial frequency response is exhibited by the material. Another type of photopolymer is the one comprising polymethylmethacrylate (PMMA) as the base (17–19). This type of photopolymer is primarily sensitive to ultraviolet light but its sensitivity could be extended to some region of the visible spectrum with sensitizers as p-benzoquinone. Diffraction efficiencies of about 70 percent can be obtained with exposures of 1 to $8 \, \text{J/cm}^2$. Problems of irreproducibility of results have been found. More recently mixtures of photopolymers containing PMMA have been studied (20–22) obtaining refractive index changes of about 10^{-2} by photoinduced polymerization.

Another approach in the fabrication of photopolymers was done at RCA laboratories (23–25). They used alfa-diketones in castable polymeric hosts like acrylics, polyesters, and epoxies. Among the alfa-diketones were benzil and camphorquinone. Diffraction efficiencies of about 70 percent have been obtained with high exposures of about $240 \, \text{J/cm}^2$, and self-development occurs slowly.

Other studies comprising polymers for holographic storage have been done. Several different materials such as optical cements (26, 27), multicomponent photopolymer systems (28), poly methyl alfa-cyanocrylates with parabenzoquinone (29), and materials that exhibit four center-type photopolymerization (30) have been used. Two polymeric materials that were applied in transient polarization holography are PMMA films in which methyl red was introduced (31) and polyvinyl alcohol (32) with methyl orange as sensitizer. Diffraction efficiencies of about 35 percent were obtained at a total intensity of about 700 mW/cm². More recently photopolymers were used in the field of optical phase conjugation (33) and holographic interferometry (34).

The photosensitive mixture

Among the holographic desired characteristics of the photosensitive medium is the ease of fabrication. The mixture proposed in this report is easy to handle and no control of room temperature and relative humidity should be done if their values fall in the normal range (temperature 20 to 28°C, relative humidity 40 to 70 percent). The mixture consists of a monomer (acrylamide), a promoter (TEA), a binder (polyvinyl alcohol, PVA), and a sensitizer (MB). The polyvinyl alcohol is dissolved in a mixture of water and alcohol and then the remaining components are added to form the mixture (in solution), which is insensitive to room light conditions (frosted bulbs). See Table 10.1 for a description of a typical composition. The amount of solution that is poured over a leveled substrate (glass) is a function of the desired thickness of the sensitive layer. After the deposition a cover could be placed above the plate to prevent particles of dust from falling on the film until it dries, about 8 hours later. This drying period is a function of the amount of materials used in the photosensitive solution. The size of the plates used in the experiment was about 5×5 cm; however, after the drying period it was found that about 1 cm near the edges of the plates was useless to record information because a slight bump was present. The thickness of the plate in its central region was about 40 μm and in the periphery about 60 to 70 μm. Just the area in the central part was used in the experiments described hereafter.

Table 10.1. Typical composition to prepare the photosensitive mixture

Solution I	
Polyvinyl alcohol	7.5 g
H$_2$O	100 ml
Alcohol	65 ml
Solution II	
Solution I	15 ml
Acrylamide	0.3 g
Triethalonamine	0.1 ml
Methylene blue	3.0 mg

Chemicals are dissolved in distilled water. The glass substrates should be clean to assure good adhesion of the polymer.

Dye uses as a sensitizer was MB, which has an absorption band in the red part of the spectrum, so an He–Ne laser (632.8 nm) used to record the information in the plates. This is an advantage because of the availability and reliability of this type of laser. Transmittance of the plates as a function of the exposure was measured by sending a collimated red beam (632.8 nm) to the plate. To avoid the formation of volume gratings during the exposure time, due to substrate reflections, a prism was placed in contact with the glass plate during the transmittance measurements. An index matching liquid filled the gap between them. The parameter of this experiment was chosen to be the beam recording power. Figure 10.1 shows the transmittance behavior as a function of the recording time. It is noticed that maximum transmittance is a function of the recording beam power. The action of light on the photosensitive material is twofold; it bleaches the dye and promotes polymerization. The result is a medium with spatial changes in refractive index and absorption. If only absorption were present after irradiation of the polymer, the desirable "gamma" of the exposed emulsion to achieve holographic linear recording (35) would need to have a value of 2. From the experimental data shown in Figure 10.1 we can see that the

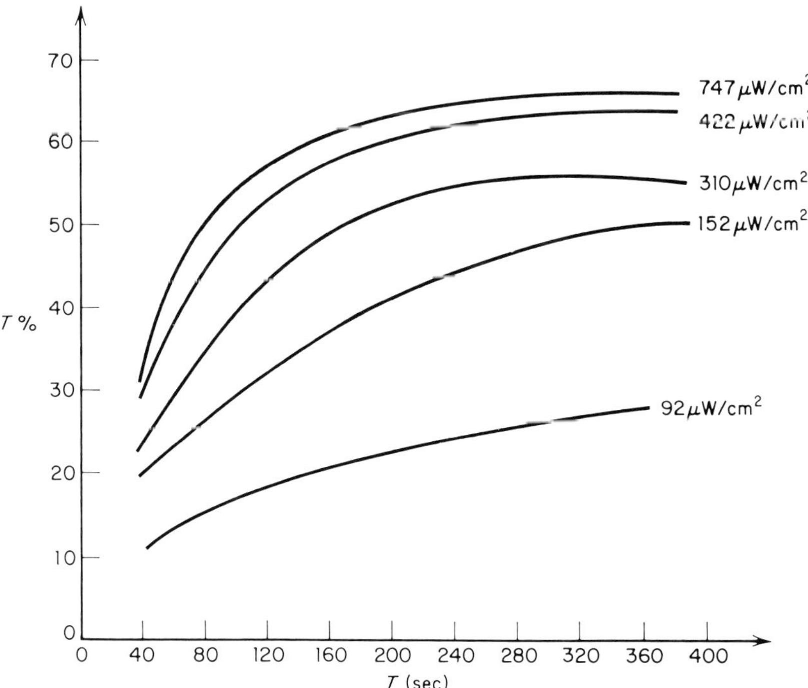

Figure 10.1. The curve that could present the highest value of gamma is the one that was induced by the beam with the highest power (747 μW/cm^2).

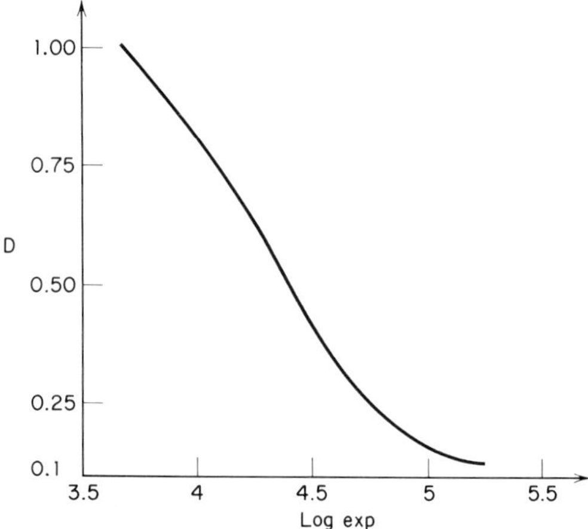

Figure 10.2. Calculating the optical density as a function of logarithmic exposure.

curve that could present the highest value of gamma is the one that was induced by the beam with the highest power ($747\,\mu\mathrm{W/cm}^2$). Therefore, by calculating the optical density as a function of logarithmic exposure, Figure 10.2 is obtained. The value of the gamma in the linear region is about -0.8, which is far from the ideal value. It should be remembered, however, that absorption modulation is not the only effect of irradiation. Change in refractive index is also present.

Interferometric studies

To characterize the photosensitive medium several interferometric experiments were performed. Interference gratings were recorded and values of diffraction efficiency and exposure were measured. Spatial frequency of the interference pattern was held fixed at $100\,\mathrm{l/mm}$. To measure the diffraciton efficiency a He–Ne laser different from the recording one was used. Using an electronic shutter, light from the writing laser was obstructed while the light from the reading laser was permitted to pass. This measurement did not affect the recorded pattern because the time needed to make the measurement was extremely short. Diffraction efficiency as a function of the recording time is shown in Figure 10.3. The beam ratio has a value of 1. The parameter in Figure 10.3 is the combined power of the recording beams. Most of the curves in this figure reach a maximum, the value being a function of the power of the recording beams. To emphasize this behavior the maximum value of the diffraction efficiency has been plotted as a function of the writing beams' power and is shown in Figure 10.4. Unfortunately the lack of a laser with more power prevented us from studying the material when recording beams with more than $235\,\mu\mathrm{W/cm}^2$

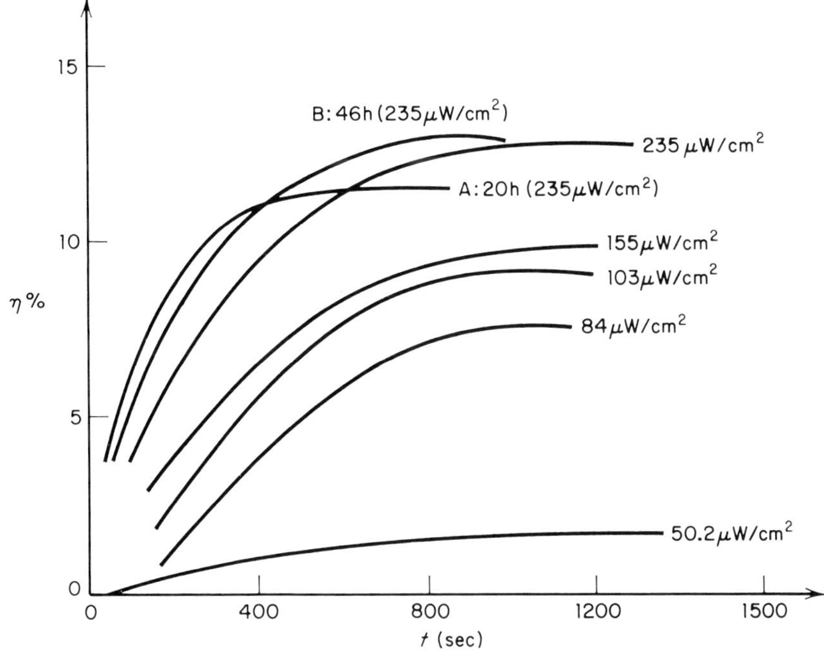

Figure 10.3. Diffraction efficiency as a function of the recording time.

were present. One possible explanation about the dependence of the change in refractive index on the irradiance of the exposing light (shown in Figure 10.4) is that the average molecular weight of the polymer molecules depends on the free radical concentration, which in turn depends on the irradiance (4).

As has been mentioned in the introductory section some of the polymeric photosensitive mixtures cited in the literature degrade in the course of time. To know the response of the photosensitive layers as the elapsed time grows from the moment of fabrication, two layers were used to record an interference pattern. One of the films was exposed after 20 hours of fabrication (plate A) and the other one after 46 hours (plate B). Results can be seen in Figure 10.3. The slope of the plate A is more pronounced than that of plate B, meaning that sensitivity is higher for plate A. However, the maximum obtained diffraction efficiency is for plate B, therefore, the maximum refractive index change is higher in plate B. The combined power of the writing beams $(235\,\mu\text{W/cm}^2)$ and the spatial frequency of the interference pattern were held constant during this part of the experiment.

One of the parameters that should be considered at the recording step of a hologram is the beam ratio (36). This was investigated for the material under study. Diffraction efficiency as a function of exposure time was

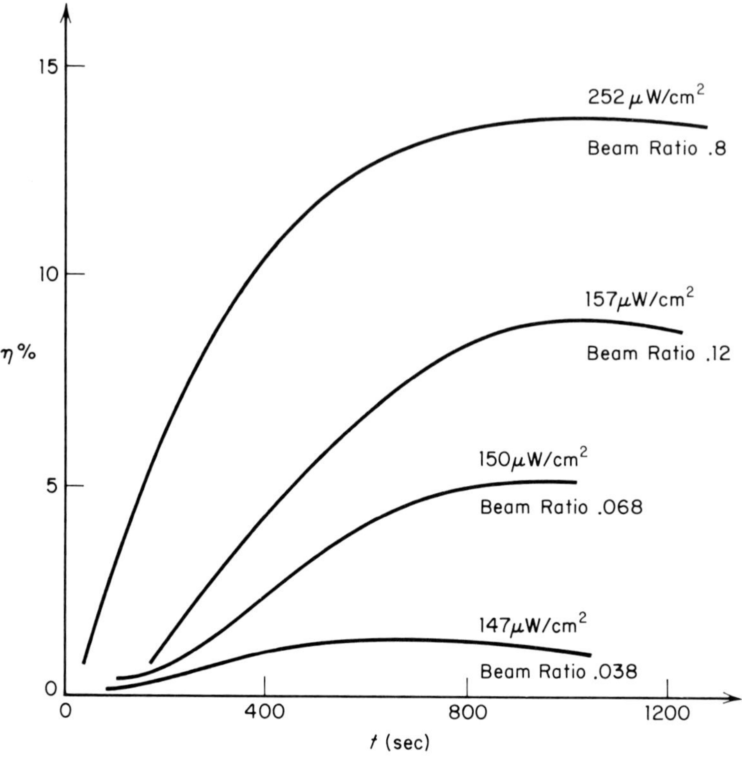

Figure 10.4. The diffraction efficiency plotted as a function of the writing beams' power.

measured for an interference grating having a spatial frequency of 100 l/mm. The beam ratio was varied by inserting neutral density filters in one arm of the configuration. Results shown in Figure 10.4 indicate that the maximum diffraction efficiency attained for each value of the beam ratio is different, meaning that the modulation was also different. It should be remembered, however, that in this type of material the change in refractive index is a function of the power of the writing beam. Therefore, the decrease in the maximum attainable diffraction efficiency for each curve (Figure 10.4) is due to two factors; the change in the beam ratio and the decrease of the total writing beams' power.

The recording of interference gratings or holograms assumes some spatial modulation of the photosensitive medium. For photographic plates a spatial change in absorption is present after irradiation and development. In the photopolymers cited in the introductory part of this chapter, the modulation could be present as a form of variations in the index of refraction (volume modulation) or variations in the thickness of the layer (surface modulation). To investigate if the diffraction phenomena given by the irradiated polymer

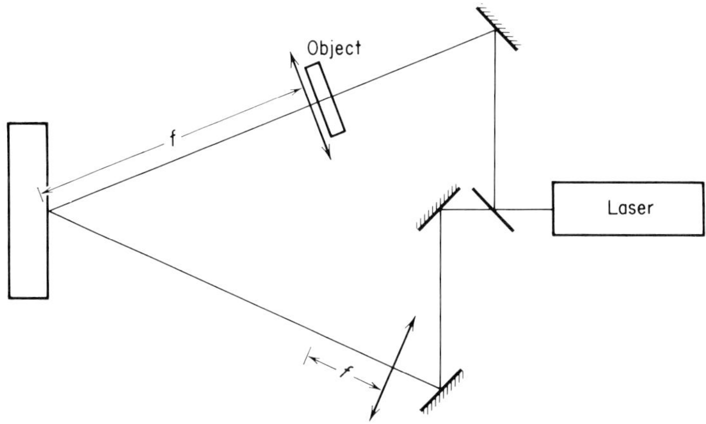

Figure 10.5. Setup to differentiate recording by volume or surface effects.

used in the present work was due to a volume effect or to a surface effect, two tests were performed. In one an interference grating was recorded in the usual way. Then a plate of glass was placed in contact with the photopolymer and a matching liquid was poured between them with the result that a very slight reduction in the intensity of the first diffracted order was noticed. The second test was the recording of an interference grating while the photopolymer was inside a liquid gate. After the exposure the grating was reilluminated and the diffracted orders were seen. With these results it is possible to conclude that the modulation is present in the bulk or volume of the photosensitive layer (Fig. 10.5).

Dye behavior

As mentioned earlier the dye used to sensitize the polymer was MB because it presents high absorption at 632.8 nm. Several interesting phenomena were noticed after the recording of an interference pattern. In the exposed area the dye was partially or totally bleached; however, if the plate was left in a light-tight container after a few hours the dye recovered itself and presented its former color. Moreover, if the irradiated area were investigated with a red-colored beam the diffracted orders were seen. This implies that the dye passed by a transition to the leuco-form or the excited state, and then it decayed to the ground state. The polymerization of the exposed parts remained, however, even though the dye recovered when in the dark.

Although the photosensitive medium described here can be used in real-time holography (i.e., reconstruction can be made just after exposure), it is interesting to investigate a fixing method, that is, a process by means of which the dye could be rendered colorless even if the plate is stored in dark room conditions after the exposure time. Jenny (4, 5) mentioned three methods to fix irradiated polymers. In one of them at the fabrication time a substance is introduced in the polymer, so that after the recording time ultraviolet light is used to irradiate the polymer, and the dye is reduced temporarily to its colorless form. The use of this substance, however, gives undesired results, like lower diffraction efficiencies and high scattering. Another method supposes a flash-exposure of the irradiated plate. A xenon flash lamp filtered to transmit wavelengths that lie within the absorption band of the dye is used. In this way the remaining monomer is polymerized at a faster rate than the one presented during the holographic exposure, and the dye molecules transform to the leuco-form. The last fixing technique, thermal fixing, is permanent and involves keeping the hologram in the dark until the catalyst is completely deactivated.

A technique involving some of the steps mentioned previously was developed to fix the recorded hologram. This technique comprises the use of a mercury lamp (Osram Hg 110). After the recording of the hologram, the plate is placed in close contact with the mercury lamp. A 20-min period is enough time to render colorless the dye and terminate polymerization. This process seems effective because the hologram can be reconstructed with little noise. Holograms have been recorded and fixed, and then they were stored in a light-tight box for about 3 months. No recovery of the dye was noticed.

Applications

In this section three applications are mentioned, namely, copy of a computer-generated hologram (CGH), the recording of a Fourier hologram, and multiple imaging by holography.

The use of a CGH is increasing in fields such as optical testing (37), pattern recognition (38), and other fields (39). The CGH is calculated with the aid of a computer and then it is displayed in a screen or in hardcopy. After this a reduction of the pattern is performed. In the present work a Fourier hologram was calculated and displayed on a hardcopy. Then a reduction of it was done by means of photography. The film used was Kodak 2415. The reduction of the CGH was placed in close contact with the photopolymer. An index-matching liquid was poured between them and then the set was illuminated with red light from a He–Ne laser for about 1.5 min. The beam power was about $1.7\,\text{mW/cm}^2$. Figure 10.6 shows three photographs: (a) the digitalized input scene, after it was introduced to the computer by means of a television camera, (b) the Fourier hologram optical reconstruction given by the CGH in Kodak film, and (c) the Fourier hologram reconstruction of the CGH contact printed in the polymer. An $f/12$ lens was used in the reconstruction. In photographs (Fig. 10.6b and

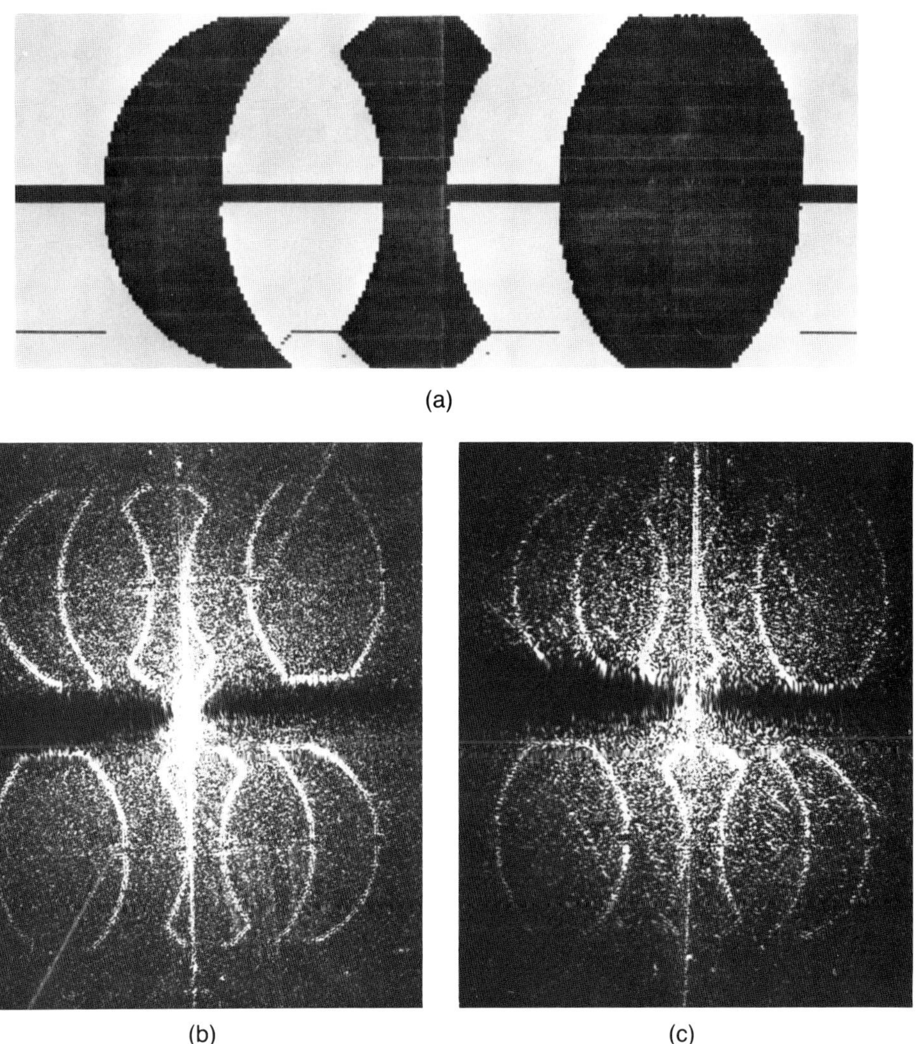

(a)

(b)   (c)

Figure 10.6. Computer-generated hologram.

10.6c) it is possible to notice the two images given by the Fourier hologram. The zero order was blocked by means of a black paper. It should be noted that the output of the Fast Fourier Transform algorithm has a too wide dynamic range for quantization in some levels giving as a result the differentiation of the images seen in Figure 10.6.

The recording of holograms by means of the photopolymer was done using a Fourier hologram configuration (40). The object used was a test chart. The trajectories of the reference and object beams were disposed at

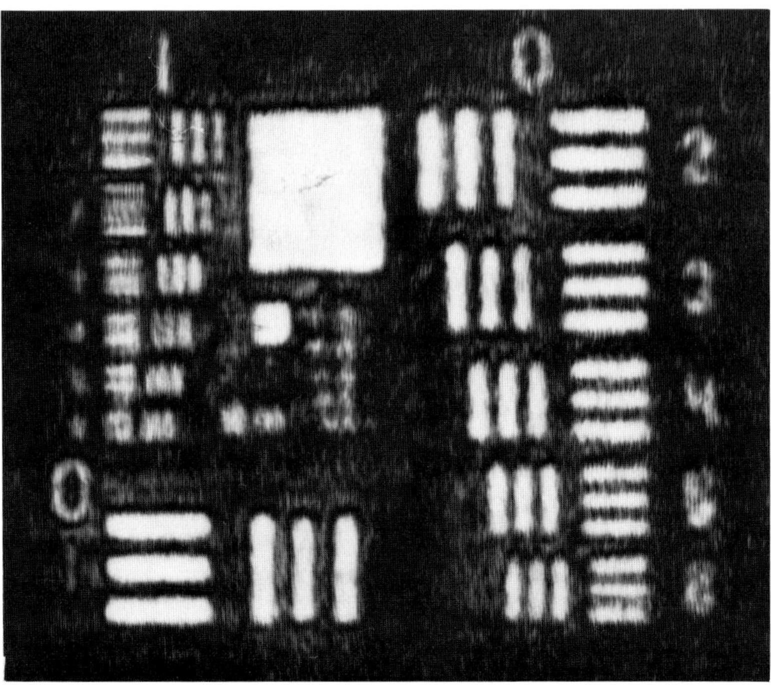

Figure 10.7. Hologram of a test chart recorded on photopolymer

an angle so as to give a mean spatial frequency of about 115 l/mm. On reconstruction with a plane wavefront, in real-time, the images given by the hologram were bright. A photograph of one of the images is shown in Figure 10.7.

The formation of multiple images from a single one can be done by different methods, such as the step-and-repeat camera process, by means of a fly's eye lens, or by an array of pinholes. Each of these methods has its advantages. Recently a method based on holography has been described, the multiple imaging by lensless transform holography (41). Basically in this method several reference beams are generated by inserting a grating in the reference beam path. The grating simulates an array of symmetrical point sources in the plane containing the reference point source. This idea was implemented by placing a ronchi grating in the reference beam path of a Fourier hologram configuration. The object was a small section of a test chart. The photopolymer was used to record the interference pattern and in reconstruction, in real-time, the image seen was that of Figure 10.8. It was possible to notice three images, each with a different intensity. This fact can be explained by recalling that when the ronchi grating was inserted in the reference beam path, zero order intensity was higher than that presented by other orders. Thus, the modulation induced in the polymer by the

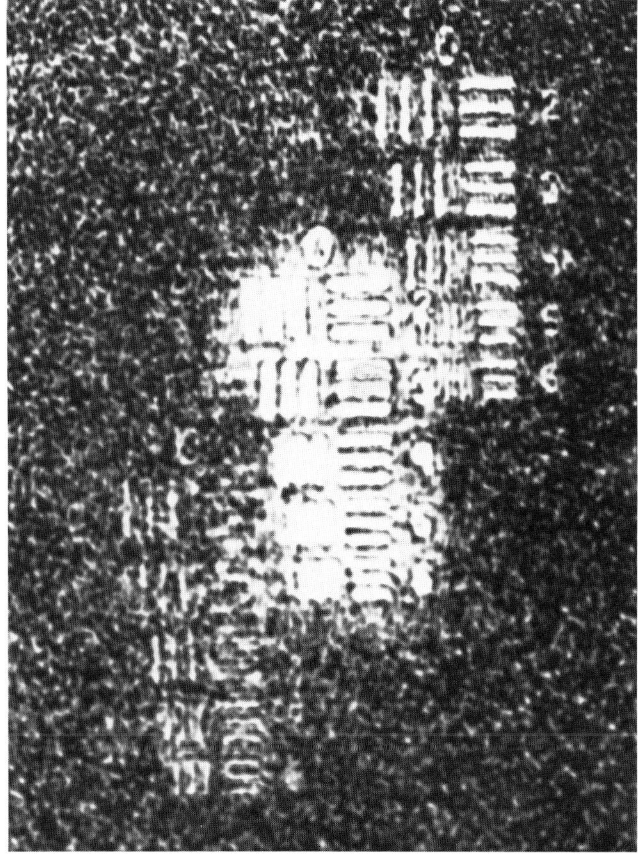

Figure 10.8. Hologram of multiple images on photopolymer.

interference between the object beam and each of the diffracted orders will be different, giving at the reconstruction step images with different brightness.

Conclusions

The feasibility of using dry acrylamide–polyvinyl alcohol-based films as holographic material has been shown. Among its characteristics are ease of fabrication and the attainment of diffraction efficiencies of about 10 percent with exposures of about 94 mJ/cm^2. The recorded holograms can be read instantaneously because the polymer is self-developing, but if necessary, a fixing technique has been proposed. It is desirable to use lasers having output powers in the order of 100 mW because it has been found that higher diffraction efficiencies can be attained if the writing density power is high.

Applications in the field of Fourier holography and copy of CGH by the contact method have been shown. Further results concerning the improvement of the sensitivity by varying the parameters that compose the polymeric mixture and details about the mechanism involving the interaction of light with the polymer will be presented later.

Acknowledgments

Thanks are given to the Instituto de Investigaciones Medicas, Universidad de Guanajuato, Mexico, for supplying some of the materials used in this work. Also thanks are given to M. Servin-Guirado for the calculation of the CGH.

References

1. Smith, H. M., Holographic recording materials. *Topics in Applied Physics*, Vol. 20, Springer Verlag, Berlin, Heidelberg, N.Y. (1977).
2. Close, D. H., Jacobson, A. D., Margerum, J. D., Brault, R. G., and McClung, F. J. *Applied Physics Letters* **14**, 159 (1969).
3. Sugawara, S., Murase, K., and Kitayama, T. *Applied Optics* **14**, 378 (1975).
4. Jenny, J. A. *J. of the Opt. Soc. Am.* **60**, 1155 (1970).
5. Jenny, J. A. *JOSA* **61**, 1116 (1971).
6. Jenny, J. A. *Applied Optics* **11**, 1371 (1972).
7. Van Renesse, R. L. *Optics and Laser Technology* **4**, 24 (1972).
8. Sadlej, N. *Optics and Laser Technology* **5**, 230 (1973).
9. Jeudy, M. J., and Robillard, J. J. *Optics Communications* **13**, 25 (1975).
10. Sadlej, N. and Smolinska, B. *Optics and Laser Technology* **7**, 175 (1974).
11. Wopschal, R. H. *J. of the Opt. Soc. of Am.* **61**, 649A (1971).
12. Colburn, W. S., and Haines, K. A. *Applied Optics* **10**, 1636 (1971).
13. Booth, B. L. *Applied Optics* **11**, 2994 (1972).
14. Booth, B. L. *Applied Optics* **14**, 593 (1975).
15. Booth, B. L. *Proc. SPIE* **123**, 38 (1977).
16. Wopschall, R. E., and Pampalone, T. R. *Applied Optics* **11**, 2096 (1972).
17. Tomlinson, W. J., Kaminow, I. P., Chandros, E. A., Fork, R. L., and Silfvast, W. T. *Applied Physics Letters* **16**, 486 (1970).
18. Moram, J. M., and Kaminow, I. P. *Applied Optics* **12**, 1964 (1973).
19. Bowden, M. T., Chandros, E. A., and Kaminow, I. P. *Applied Optics* **13**, 112 (1974).
20. Franke, H., Festl, H. G., and Kratzig, E. *Colloidal and Polymer Science* **262**, 213 (1984).
21. Kpietz, M., Lechner, M. D., and Steinmeier, D. G. *Makromol. Chem. Rapid Comm.* **4**, 113 (1983).
22. Hillmar, F. *Applied Optics* **23**, 2729 (1984).
23. Bloom, A., Bartolini, R. A., and Ross, D. L. *Applied Physics Letters* **24**, 612 (1974).

24. Bartolini, R. A., Bloom, A., and Weakliem, H. A. *Applied Optics* **15,** 1261 (1976).
25. Bloom, A., Bartolini, R. A., Hung, P. L. K., and Ross, D. L. *Applied Physics Letters* **29,** 483 (1976).
26. Files, V. *Applied Physics Letters* **19,** 451 (1971).
27. Forshaw, M. R. B. *Optics and Laser Technology* **7,** 28 (1974).
28. Tomlinson, W. J., Chandros, E. A., Weber, H. P., and Aumiller, G. D. *Applied Optics* **15,** 534 (1976).
29. Friesem, A. A., Ray-Nov, Z., and Reich, S. *Applied Optics* **16,** 427 (1977).
30. Mizuno, T., and Hattori, S. *J. of the Opt. Soc. of Am.* **67,** 1651 (1977).
31. Todorov, T., Tomova, N., and Nikolova, L. *Optics Communications* **47,** 123 (1983).
32. Todorov, T., Nikolova, L., and Tomova, N. *Applied Optics* **23,** 4309 (1984).
33. Dennis, W. M., Blau, W., and Bradley, D. J. *Optical Engineering* **25,** 538 (1986).
34. Oliva, J., Mateos, F., and Pastor, C. *Optica pura y aplicada* **18,** 193 (1985).
35. Bachman, P. L. Silver halide photography. In *Handbook of optical holography* (H. J. Caulfield ed.), Academic Press, N.Y. (1979).
36. Friesem, A. A., Kozma, A., and Adams, F. *Applied Optics* **6,** 851 (1967).
37. Loomis, J. S. *Optical Engineering* **19,** 679 (1980).
38. Casasent, D. *Optical Engineering* **24,** 724 (1985).
39. Proceedings SPIE **437,** International conference on computed generated holography, p. 19 (1984).
40. Lowenthal, S., and Belvaux, Yv. *Revue d'Optique* **46,** 1 (1967).
41. Mehta, P. C., Bhan, C., and Hradaynath, R. *J. Optics (Paris)* **10,** 133 (1979).

11
Variable index materials for optical data processing

JEAN J. ROBILLARD

Since the discovery of holography by Dennis Gabor in the mid-1960s (1) scientists and engineers have been aware of its potential, not only for three-dimensional displays but also, and more importantly, for optical data processing, data storage, and nondestructive testings. The new technology did not develop as fast as could be expected, mainly because of the lack of proper recording media. The ideal material for hologram recording should provide a variation of index of refraction, Δn, as response to light. For high efficiency the material should be perfectly transparent and the variations of index should not be accompanied by absorption or diffusion. This appears to be incompatible with any recording technique as the diffracted light forming the hologram should transfer some energy to the material to cause the index to change and, therefore, absorption should occur. A simple solution to the problem was provided by the induced spectral sensitization technique (ISST) (2, 3) whereby a spectral sensitizer for the photosensitive process responsible for the index variation was created temporarily in the material during the recording process and thereafter withdrawn to restore complete transparency. This was made possible by the use of a photo-chrome as a sensitizer. An ultraviolet (UV) source would trigger the sensitization in a material that is normally transparent for the recording light of the laser. Absorption would then occur during the recording of the hologram to provide the necessary variations of index and, upon completion, the UV illumination would be turned off, the sensitizer would disappear, and the material would return to its original transparency. In that the hologram is being recorded with pure index variations, its efficiency is optimum (4). The quality of the hologram is not the only advantage of the method: it is a real-time recording and the use of two wavelengths to create the variation of index provides a nonlinear optical material responding to two signals that can be combined for data processing. For holography, three alternatives are then possible: (1) a UV hologram with overall illumination in the visible, (2) a hologram in the visible with overall illumination in the UV, and (3) the combination of two holograms, one in UV and the other in the visible.

136

$$\left.\begin{array}{c} \lambda_1 \\ \\ \lambda_2 \end{array}\right\} \quad \text{VIM} \quad \Delta n = f(\lambda_1, \lambda_2) \quad \left\{\begin{array}{l} \lambda_1 \ \text{Hologram with } \lambda_2 \ \text{Illumination} \\ \lambda_2 \ \text{Hologram with } \lambda_1 \ \text{Illumination} \\ \lambda_1 \lambda_2 \ \text{Combination Hologram} \end{array}\right.$$

$$\left.\begin{array}{c} \lambda \\ \\ \vec{E} \end{array}\right\} \quad \text{VIM} \quad \Delta n = f(E, \lambda) \quad \left\{\begin{array}{l} \lambda \ \text{Hologram with Applied } E \\ \text{Computerized Hologram} \\ (\text{Crossbar} + \lambda \ \text{Illumination}) \\ \text{Combination } \lambda + E \end{array}\right.$$

$$\left.\begin{array}{c} \lambda \\ \\ T \end{array}\right\} \quad \text{VIM} \quad \Delta n = f(\lambda, \Delta T) \quad \left\{\begin{array}{l} \lambda \ \text{Hologram with } \Delta T \\ \text{Infrared Hologram with } \lambda \\ \text{Illumination} \\ \text{Combination } \lambda + \Delta T \end{array}\right.$$

Figure 11.1. The alternatives in the ISST technique.

The ISST technique was subsequently extended (5) to include other physical means to control the spectral sensitization such as an electric field (electrochromism) and temperature (thermochromism). The new recording materials would then allow the mixing of electrical and optical signals in one instance and thermal and optical signals in the other. The latter case would be particularly suitable to infrared holography. The various alternatives are summarized on Figure 11.1.

The variable index materials (VIM) are stable in a normal environment as the simultaneous occurrence of two kinds of activation is necessary to alter the index of refraction and, for the same reason, the hologram does not require fixing. Also, additional exposure can be provided in already exposed areas if the saturation has not been reached. New exposures can also be made in areas not previously exposed.

The index of refraction of a transparent material is related to the existence of dipoles in its electronic structure. Dipoles can be formed by single electrons associated with a positive vacancy or a hole, molecular groups with unsymmetrical charges distribution, or boundary layers with potential barrier made of opposite charges. In organic semiconductors, electrons can be trapped at vacancies, which often result from a perturbation in the periodic distribution of the molecular field and the system electron-trap provides a dipolar contribution to the index. In that case a change in refracting index can be promoted by trapping photoelectrons.

Dipolar effects due to charges distribution are generally induced by delocalization of electron orbitals or change in symmetry due to photo-isomerization. It then leads to a bistable situation corresponding to two isomers with different indices. Such mechanism can be initiated in a variety of ways involving a single material that is an organic semiconductor or the combination of an organic semiconductor with a photocrosslinkable or photodegradable material. The photoinitiation of a copolymerization or crosslinking between unpaired reactive groups in the monomers results in unsaturated bonds distributed in the structure of the polymer. These localized active bonds act as an electron trap, electron donors or acceptors, or delocalization sites for the electrons of double bonds in the vicinity of the sites. This mechanism called *lacunar copolymerization* (6) provides a relatively large ($\Delta n = 10^{-2}$) change in index of refraction.

If the nature of the bonds in a crosslinkable polymer is such that they could easily be promoted or dissociated under a bistable situation, as referred to earlier, the variation of indices would be reversible.

The photosensitive processes involved in the VIM using the ISST technique are photopolymerization and photocrosslinking. These processes are not reversible and can only be applied to the recording of nonrecurrent events. In photopolymerization the variation of index is due mainly to a change in density due to the association of monomer molecules into a macromolecule. In photocrosslinking, index variations are due to a change in density as well as the direct contribution of the electrons with low binding energy of the functional groups responsible for the crosslinking. This last contribution can be enhanced by electron exchanges or electron interaction with nearby sites of an organic semiconductor.

Index of refraction of a transparent material

The index of a material depends on its electronic structure, of the distribution of the electrons around the atoms of the substance, their density, bonding modes, and other factors.

One can express the contribution of each molecular component to the index of refraction of the material by the molecular refraction R_M:

$$R_M = \frac{n^2 - 1}{n^2 + 2} \frac{M}{\rho} \tag{11.1}$$

where

n = the index of refraction

M = the molecular weight

ρ = the density of the material

In the organic material, each atom, atomic group, or chemical bond provides a well-defined contribution to the value of the index as can be seen on Table 11.1.

If the material does not contain permanent dipole in the electronic structure and, for larger wavelength, the Maxwell relation:

$$n^2 = \varepsilon$$

Table 11.1. Molar refraction (R_M)

		Bonds
H: 1.1	H_2O: 3.75	
C: 2.42	CH_2: 4.61	C=C: 4.16
Cl: 5.42	C_6H_6: 26.18	C=N: 5.45
Br: 8.96	CCl_4: 26.51	C≡C: 6.40
O: 1.52	C_2H_5OH: 12.78	C=O: 1.51

where n is the index of refraction and ε the dielectric constant, is applicable, and the molecular refraction is then equal to the molecular polarization.

Methods for calculation of n

The value of R_M that characterizes the polarizability of the molecule can be obtained from its electronic structure using various approximation methods.

Method of Mulliken

The method of Mulliken (7) is based on the electronic structure of the atomic bonds present in the molecule. The molecular refraction R_M is equal to:

$$R_M = 1.705l + 1.208m + 2.942n + E$$

where:

l = the number of C–H bonds

m = the number of σ_{C-C} bonds

n = the number of π_{C-C} bonds

E = represents the influence of the delocalization of mobile electrons in the conjugated molecules.

Method of Palit

The method of Palit (8) involves essentially the number of electrons in each atom of the molecules and their bonding mode in the atom (in contrast with the method of Mulliken, which utilizes the bonding between atoms).

Classical theory

The classical theory (9, 10) gives:

$$R_M = \frac{Ne^2}{3\pi mC^2} \sum \frac{f_j}{v_j - v} \tag{11.2}$$

where:

f_j = the oscillator strength of the absorption band whose maximum frequency is v_j

v_j = the frequency at which the material is transparent.

Variations of n

Generally speaking one can say that the index of refraction of a material corresponds to the density of electrons per unit of volume and to the ability of these electrons to oscillate. It is, therefore, important to consider the nature of the atomic or molecular bonds and the electron orbitals involved in such bonds.

To modify the index of refraction of a material, one can introduce or withdraw atoms or atomic groups through a chemical reaction. One can also promote or break certain atomic bonds in the molecules of the material. It is also possible to increase or decrease the electron density in establishing new orbitals corresponding to free-energy levels in the molecules. Organic semiconductors provide physical means to modify the local distribution of electron density thus creating dipole action contributing to the index. Plastic deformations resulting in local change in density will also produce a change in refractive index.

The ability to induce a change in refractive index as a result of light absorption will be related to the modification of one of the parameters listed previously through energy transfer from the absorbed photon to the electronic structure of the material.

Photo-induced variation of refractive index
in transparent materials

Photopolymerization

Photopolymerization is a process by which macromolecules with high molecular weight are formed from smaller molecules called monomers by the action of light in the presence of suitable reaction initiators. The absorption of radiation by the initiator produces an active molecule that can attach itself to a monomer molecule to form another reactive molecule of greater complexity and higher molecular weight. The new reactive molecule can now attach itself to another monomer molecule to form still another larger, heavier, reactive molecule. The process thus continues until the reaction stops. In the absence of other causes the termination would occur when all monomer molecules would have been consumed. Unfortunately premature termination often happens due to the presence of impurities or oxygen reacting with one of the intermediate species, thus breaking the chain reaction process. If the monomer molecule contains one functional group the resulting polymer is made of linear chains of molecules. Polyfunctional monomers provide crosslinked polymers resulting in a three-dimensional network with higher packing density.

Reaction initiators are free radicals that can be produced in two different ways: (1) photodissociation of an organic or inorganic compound or (2) photosensitized charge transfer and initiation of an ion radical.

Photodissociation
Many compounds can produce free radicals by absorption of light, generally in the UV end of spectrum. Examples are benzophenone, peroxides, diazonium salts, carbon tetrabromide, and zinc oxide.

A particular system for the generation of free radicals is the combination of a photoreducible dye (i.e., methylene blue) with an electron donor (i.e., triethanolamine). This system is of great interest because it is initiated by radiation in the visible.

Charge transfer
The formation of ion radicals by charge transfer requires the formation of a charge transfer complex resulting from the association of an electron donor and an electron acceptor. The absorption of light by the complex provides species capable of initiating the polymerization. Examples of such couples are: N-vinylcarbazole (donor) and sodium chloraurate (acceptor) or triethylaluminum (donor) and methyl methacrylate (acceptor).

One of the great advantages of photopolymerization is the fact that it is a chain reaction and, therefore, involves amplification. Indeed amplification factors of 10^5 are possible, which make the process almost as sensitive as the silver–halide process (10^9).

The addition of many monomer units resulting in a compact network brings about a change in many physical properties including often a drastic change in refractive index. This is primarily due to (1) a change in electron density that follows the increase in molecular weight, (2) the disappearance of the weakly bonded electrons of the reactive groups in the monomer that contributed to the index of refraction by their dipolar momentum but are now bounded to the next monomer molecule to form the polymer, and (3) the damping conditions that vary as the material changes into a more condensed state (i.e., liquid monomer to solid polymer). Because of the amplification involved in the chain reaction, the physical changes resulting from photopolymerization can never be reversible.

Photocrosslinking

Photocrosslinking results from the bonding of nonsaturated functional groups still existing on the macromolecules of a polymer. The bonding mechanism is similar to photopolymerization and the reaction initiators are identical. The major difference is that the change in densities is minimal and there is little or no amplification. The change in refractive index in this case is mainly due to the bonding electrons of the reactive groups that abandon their dipolar contribution. But the lack of amplification can turn into an advantage if the crosslinking bonds can be easily broken, thus leading to reversible physical changes. Reversible changes of index of refraction are of major interest for optical data processing. Certain free radical initiators can promote both the bonding or the breaking of crosslinks in the polymer. This is particularly true for charge transfer complexes involving organometallic compounds. Also, bistable structures can be created by the presence of an organic semiconductor acting as a charge transfer complex but under a different mechanism (photoconduction or field effect).

Photodegradation

Photodegradation is the reverse of photocrosslinking. It involves the photosensitized breaking of crosslinking bonds. This phenomenon occurs naturally in most of the plastic materials exposed to daylight. It is generally a slow process with very low efficiency. But certain structures can be made photodegradable with acceptable energies of activation. This is particularly the case for structures built up under the conditions set forth in the preceding section.

Lacunar copolymerization (6)

In a polymerization process, when all the molecules of monomer are identical, their polymerization is called *homopolymerization*. If the molecules are of two different species it is called *copolymerization*. In copolymerization the monomers can have the same number of functional (reactive) groups, which become saturated during polymerization. If this number is different, the polymer will contain unsaturated groups that will create local perturbations in the electronic structure of the polymer. The presence of these groups forms active centers inducing a delocalization of electron orbitals in the neighbouring molecules and they create energy levels similar to those existing in semiconductors. Such a situation exists each time a functional group of a monomer molecule does not encounter its homologue in another molecule. Figure 11.2 shows the crosslinking of two linear polymer molecules with (Fig. 11.2a) paired functional groups and (Fig. 11.2b) unpaired functional groups. ○ represents a nonsaturated functional group and ☐ a crosslink between the molecules of polymer. In Figure 11.2c, not only unpaired groups exist but some of the available groups are too far away from each other to react and form the crosslink. Figure 11.2d corresponds to the general case including Figure 11.2b and c.

The active centers distributed in the polymer will allow electron trapping and the resulting change in index of refraction.

The promotion of electrons, which can be trapped in the active centers of the polymer, will result from (Fig. 11.3):

1. Metal ions existing in the copolymerization solution and remaining trapped in the polymer
2. Organometallic groups in the crosslinks
3. Organometallic groups as part of the functional group itself
4. Organometallic groups existing in the linear backbone of the crosslinking polymer molecules

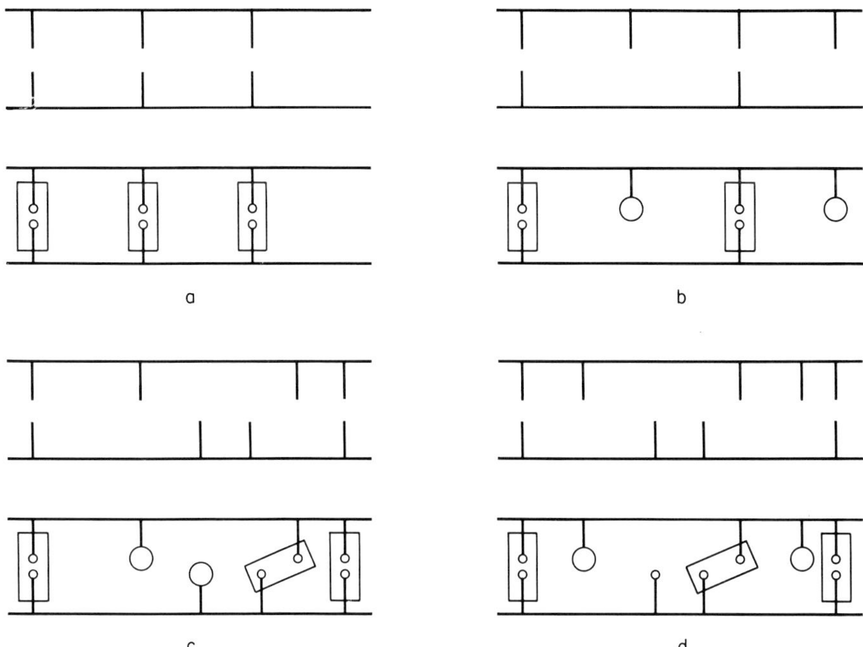

Figure 11.2. Crosslinking of two linear polymer molecules with paired functional groups and unpaired functional groups.

In all alternatives represented on Figure 11.3 the generation of an electron contributing to a change in index of refraction will result from the absorption of a photon by the metal ion, an organometallic molecule or an organic dye.

The wavelength of the exciting radiation will be related to the absorption band of the system.

Organic semiconductors

The interest of organic semiconductor as VIM is related to the electronic properties of the semiconductor and their interaction with the optical properties. If the index variation is to be induced by light, then photo-conductivity will be involved and important parameters will be the photo-current, the mobility of the charge carriers, their diffusion time, and the photodielectric effect.

Photocurrent

The variation of conductivity of the semiconductor when it is exposed to light is given by (11):

$$\Delta\sigma = e(\Delta n \cdot \mu_e + \Delta p \cdot \mu_h) \tag{11.3}$$

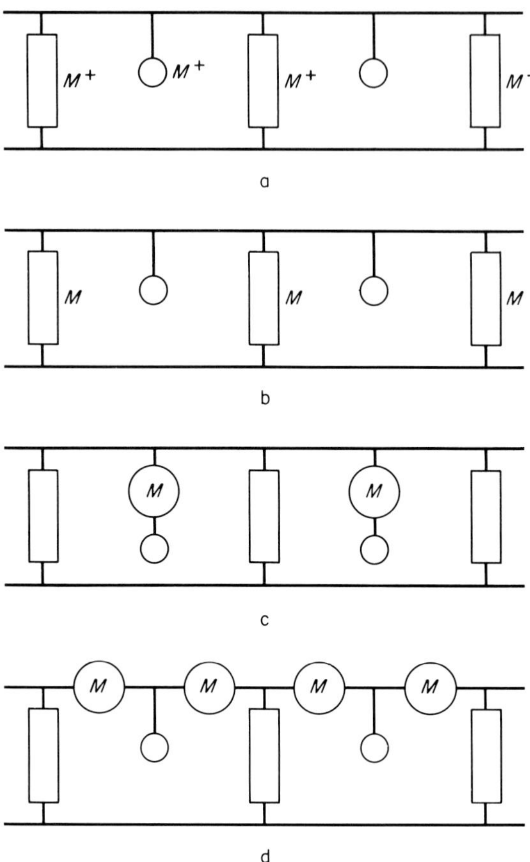

Figure 11.3. Alternatives for electron generation in crosslinked molecules.

where

e = electron charge

Δn = variation of the number of electrons

Δp = variation of the number of holes

μ_e = mobility of the electrons

μ_h = mobility of the holes

A measurement of conductivity will provide directly the number of pairs electron-hole produced by the absorbed photons. This represents the variation of the number of charges in the material that is related to a variation of index of refraction.

Mobility of the carriers (12)

The drift mobility μ_d for both free (n) and trapped (n_t) carriers is such that:

$$(n + n_t)\mu_d = n\mu \tag{11.4}$$

μ being the mobility of the free carriers. μ_d is also a function of the temperature according to:

$$\mu_d = \frac{1}{N_t} \mu \exp(-E_t/kT) \tag{11.5}$$

where N_t is the density of traps and E_t the energy level associated with the traps.

Carriers diffusion time (13)

The photoconduction efficiency G is the ratio of the number of charges flowing between the electrodes per second to the number of charges created per second:

$$G = \frac{\Delta I}{e} \Delta n \tag{11.6}$$

G can also be expressed as the ratio of a free carrier τ to its transit time t:

$$G = \frac{\tau}{t} \tag{11.7}$$

The transit time t is given by:

$$t = \frac{d^2}{\mu V} \tag{11.8}$$

where d is the distance between the electrodes and V the applied voltage.

If the life time τ of the excitons is greater than the time of absorption of a photon in the layer the light will be submitted to an index variation created by itself (14). This would mean a life time $t > 10^{-12}$ sec for a 100-μ layer with an index of 1.5.

Index variations in organic semiconductors
Contribution of the free carriers.   The refractive index is given by (15, 16):

$$n^2 = \frac{c}{2}\left[\left(1 + \frac{2\sigma}{v\varepsilon^2}\right)^{1/2} + 1\right] \tag{11.9}$$

with:

$$\varepsilon = 1 + 4\pi\chi$$

$$\chi = \frac{Ne^2}{m(2\pi G)^2} \tag{11.10}$$

n = refraction index

ε = dielectric constant

σ = conductivity

N = concentration in free electrons

m = electron mass

χ = dielectric susceptibility

μ = carriers mobility

or:

$$(2n^2 - \varepsilon)^2 = \varepsilon^2 + \left(\frac{2\sigma}{v}\right)^2 \tag{11.11}$$

and, differentiating:

$$8(2n^2 - \varepsilon)n\, dn = \frac{8\sigma}{v^2}\, d\sigma + 4n^2\, d\varepsilon \tag{11.12}$$

Assuming there is only one type of carrier (electron):

$$\sigma = Ne\mu$$

and

$$d\sigma = e\mu\, dN \tag{11.13}$$

Also, according to (16):

$$d\varepsilon = 1 + 4\pi\chi$$

$$d\chi = \frac{e^2}{4\pi^2 v^2 m}\, dN$$

or:

$$d\varepsilon = \frac{e^2\, dN}{\pi v^2 m} \tag{11.14}$$

finally:

$$\frac{dn}{dN} = \frac{e^2}{2v^2 n(2n^2 - \varepsilon)\pi m}(2\pi m\mu^2 N + n^2) \tag{11.15}$$

If $\varepsilon = n^2$:

$$\frac{dn}{dN} = \frac{e^2}{2v^2 \varepsilon\pi m}(2\pi m N\mu^2 + \varepsilon) \tag{11.16}$$

The preceding relation can also be expressed as a function of the life time of the free electron τ, knowing that:

$$m = \frac{e\tau}{\mu}$$

then:

$$\frac{dn}{dN} = \frac{e\mu}{2\pi v^2 \varepsilon\tau}(2\pi e\tau\mu N + \varepsilon). \tag{11.17}$$

In Equations 11.15 to 11.17 the first term between brackets is negligible with respect to the second, which leads to the simplified expression:

$$\frac{1}{n}\Delta n \approx \frac{e^2}{2\mu v^2 \varepsilon m}\Delta N \tag{11.18}$$

with

$$e = 1.6 \times 10^{-19}\,\mathrm{c} = 4.8 \times 10^{-10}\,\mathrm{esu}$$
$$m = 5.9 \times 10^{-28}\,\mathrm{g}$$
$$v = 10^{14}\,\mathrm{Hz}$$
$$\varepsilon^{1/2} = n = 1.5$$
$$\Delta N = 10^{18}\,\mathrm{cm}^{-3}$$
$$\frac{1}{n}\Delta n = 5 \times 10^{-4}$$

or $\Delta n \sim 0.7 \times 10^{-3}$.

Contribution of trapped carriers.   We will now assume that all carriers photogenerated are trapped, which is the other extreme. The equations of Clausius Mosotti and Lorenz Lorenz (17) lead to relation:

$$\frac{n^2 - 1}{n^2 + 2} = \frac{4\pi}{3}N\alpha \tag{11.19}$$

where

$N = $ the concentration of trapped carriers (electrons)

$\alpha = $ the electron polarizability at optical frequencies

Differentiating Equation 11.19:

$$\frac{1}{n}\frac{dn}{dN} = \frac{2\pi\alpha}{9n^3}(n^2 + 2)^2 \tag{11.20}$$

with Equation 11.19 one can also write:

$$n^2 = \frac{3 + 8\pi N\alpha}{3 - 4\pi N\alpha}, \tag{11.21}$$

which introduced into Equation 11.20 leads to:

$$\frac{1}{n}\frac{dn}{dN} = \frac{18\pi\alpha}{(3 + 8\pi N\alpha)(3 - 4\pi N\alpha)} \tag{11.22}$$

Neglecting the terms in $N\alpha$ and $(N\alpha)^2$ with respect to 1, one obtains:

$$\frac{1}{n}\Delta n \approx 2\pi\alpha \cdot \Delta N \tag{11.23}$$

with $\Delta N = 10^{17}\,\mathrm{cm}^{-3}$

and

$$\alpha = (10^{-7})^3$$

$$\frac{1}{n}\Delta N \approx 6 \times 10^{-4} \quad \text{or} \quad \Delta n \approx 10^{-3}$$

These index variations are sufficient for recording phase holograms with high efficiency.

Typical organic semiconductors for index modulation
Organometallic compounds resulting from the chelation of various polymers are easy to prepare in thin transparent films. Examples are metal chelates of polyvinyl alcohol (PVA) where the metallic ion is the crosslinking agent between the chains of PVA. These materials can be sensitized with a photoreducible dye. Light in the absorption band of the reduced dye reduces the metallic ion to a lower valence that does no longer serve as crosslinking agent (18). The disappearance of the crosslinking bond produces the change in refractive index. The opposite reaction can take place by photo-oxidation of the leuco compound at a different wavelength (UV) (19), thus yielding to a true reversible process: recording in the visible, erasing in the UV.

Examples of PVA chelates are the copper chelate (20) represented on Figure 11.4. It is obtained by adding stoichiometric proportion (one Cu for 4 OH groups in two molecules of PVA) of a copper salt to a water solution of PVA.

Another VIM complex is the titanium chelate of PVA where the crosslinking ion is introduced in the form of titanium lactate (21). In both cases auramine O can be used as a sensitizer for the 488-nm line of an argon laser.

There are a number of true organic semiconductor that can lead to exploitable change in index of refraction. One example is the poly (vinylcinnamylidene acetate) that is prepared by condensation of PVA with

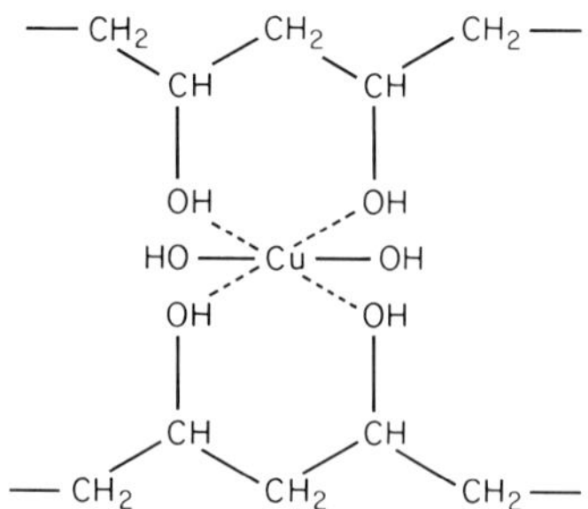

Figure 11.4. The copper chelate.

the cinnamylidene acetyl chloride. The material:

$$(CH_2\text{---}CH)_n$$
$$|$$
$$O\text{---}CO\text{---}CH{=}CH\text{---}CH{=}CH\text{---}C_6H_5$$

can be sensitized with erythrosine for the 488 nm argon line. It is soluble in a 50:50 mixture of cyclohexanone and tetrahydrofuran. But the most promising organic semiconductors for fast response variable index recording appear to be the polyacetylenes resulting from the polymerization of diacetylene monomer such as (22):

$$R_1\text{---}CH_2\text{---}C{\equiv}C\text{---}C{\equiv}C\text{---}(CH_2)_n\text{---}R_2$$

The former is being used for second harmonic generation (23) in optical data processing.

The latter, prepared by condensation of NNN′N′ tetramethylethylene diamine with bisphenol A, can be sensitized with organic dyes (erosine erythrosine) and could produce index variation $\Delta n = 1.7 \times 10^{-3}$. Other polyacetylene materials appear to be promising and are listed on Table 11.2 with the corresponding index of refraction and variation thereof. The Δn where derived from induced change in dielectric constant $\Delta\varepsilon$ assuming the relation $\varepsilon = n^2$ is applicable. The excitation was provided with a 10 μs pulse as 488 nm originating from a 5-W argon laser through a Pockel cell.

Dual spectrum response can be provided by polymers containing photochromes spirobenzopyran. These materials are normally transparent in the visible but develop an absorption band upon exposure to UV. The coloration is accompanied by a change in refractive index in the molecule itself but more importantly it induces a major change in refractive index in the macromolecular structure where it is incorporated.

The structure of the molecule is distributed in two perpendicular plans on each side of the central carbon atom. Exposition to UV breaks the C—O bond of the pyran, thereby producing the rotation of one side of the molecule, both sides becoming coplanar. The new structure develops an absorption band in the visible and light absorbed at this frequency causes the molecule to return to its original state. The photoreaction is illustrated on Figure 11.5.

The speed of the reaction in both directions varies with the substituent and can reach the microseconds (24).

The polymerization of spiropyran monomers such as (A), (B), (C), and (D) in Figure 11.6 leads to various polymers with relatively small n (5 to 12)

Table 11.2. Polymers

Polyacetylene	n	Δn
a	1.86	1.7×10^{-3}
b	1.76	5×10^{-4}
c	1.72	1.5×10^{-3}
d	1.80	2.3×10^{-3}
e	1.58	2.8×10^{-3}

(Fig. 11.7). These products respond to both UV and visible excitations providing index variation in opposite directions. The polymers resulting from the condensation of benzothiazoline type structure (C and D) are generally faster in response than their indoline homologues (A and B). The copolymerization of spiropyrans with various organic monomers should lead to materials with interesting optical properties due to the steric distortion of the π electrons systems and possible conjugation with other double bonds resulting from the ring opening of the spiropyrans. It appears that such class of polymers would be good candidates for variable index materials. The

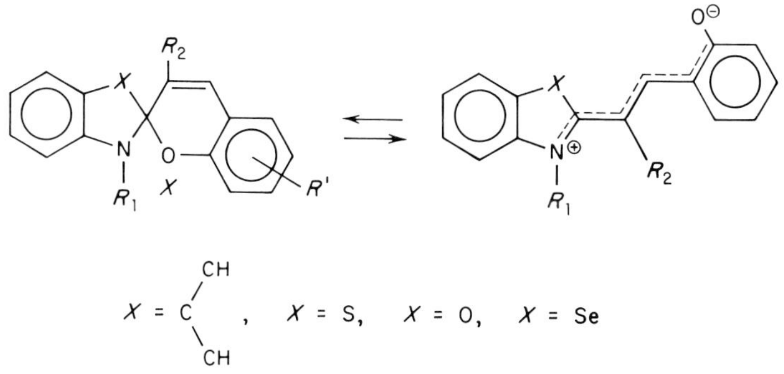

$$X = C \begin{smallmatrix} CH \\ \\ CH \end{smallmatrix}, \quad X = S, \quad X = O, \quad X = Se$$

Figure 11.5. Photoisomerization in spiropyrans.

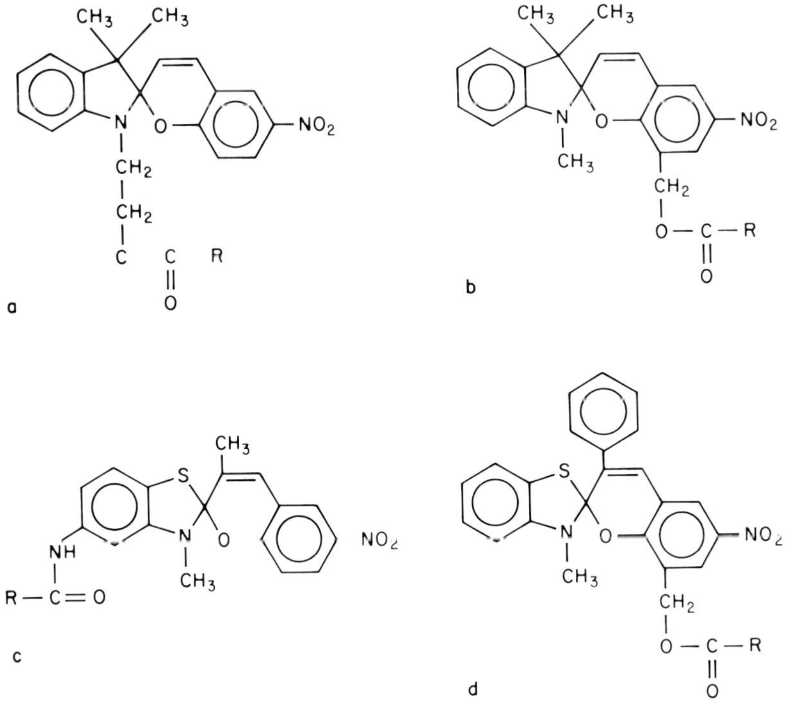

Figure 11.6. Spiropyran monomers.

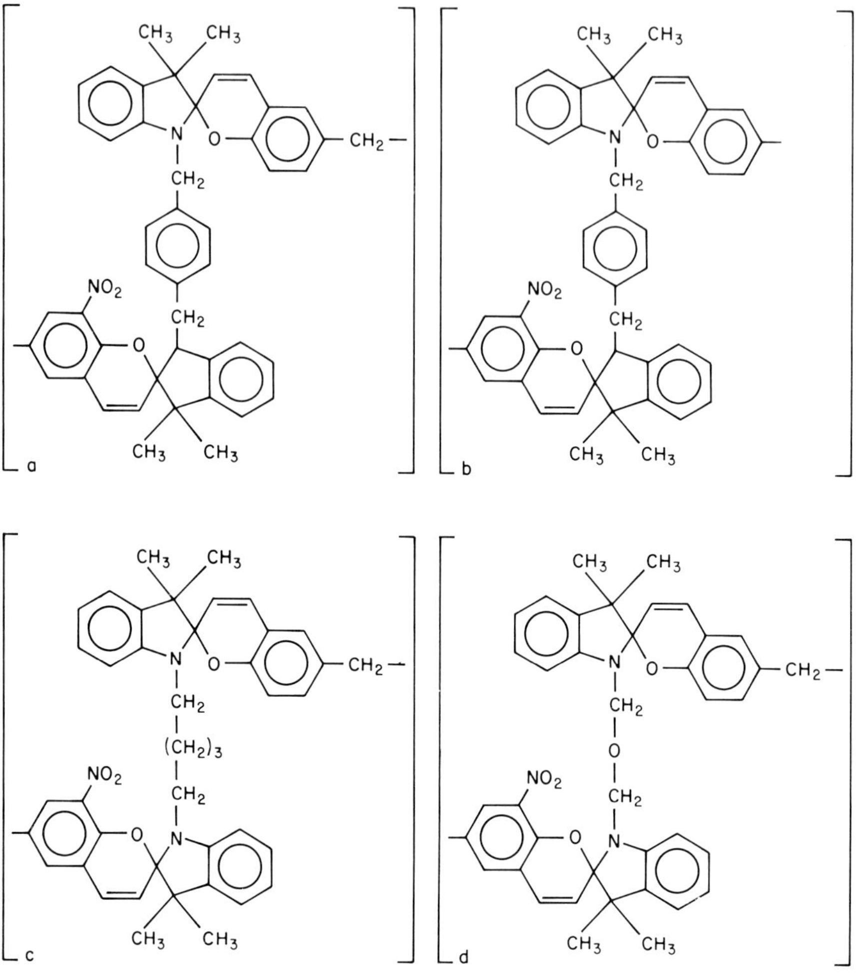

Figure 11.7. Spiropyran polymers.

synthesis and properties of polymeric spiropyrans have been studied by the group of G. Smets at the University of Louvain, Belgium (25).

Conclusion

The rapid expansion of holography and holographic techniques has created a need for phase recording materials with speed comparable to that of bleached silver halide emulsions. The advent of optical data processing and the optical computer creates even more stringent requirements on recording materials such as real-time recording, write–erase capabilities, and response time in the micro, possibly picosecond range.

Great achievement have been possible in the recording of permanent holograms through sophisticated bleaching techniques of silver halides emulsions. Also phase holograms with high efficiency can be recorded on photopolymers (25–27) and photodegradable (22) materials. These recording techniques will remain inherently slow compared with silver halide, due to the nature of the physical process involved.

Rapid access and erasing capabilities are now met by various electro-optic materials (28) but they need to be monocrystals or polycrystals with large crystallites. Crystal growing techniques and further processing for optical compatibility are complicated and expensive. The optically active area is necessarily limited in size.

Spreadable polymeric films are attractive because of the easiness of preparation. Symmetry can be introduced in their physical properties by electric field orientation or selective stretching during or after formation. Electro-optic and semiconductive properties of organic semiconductors in both crystalline and amorphous states are being discovered and new techniques are directed to enhance such properties. We are beginning to make use of electronic symmetry in molecules in the same way as it is currently done in crystals. Existing molecular structures are modified to provide ausotropy, carrier mobility, oscillation coupling, leading to a temporary or permanent modification of optical or electrical properties under the influence of various physical excitations. Quantum efficiency up to 10^{12} has been obtained tor the polymerization of certain organic semiconductors (29). Second harmonic generation has been performed in diacetylene polymer films (30). Infrared absorption has been used to modulate the two indices of a photo dichroic polymer film for visible imaging of an infrared pattern distribution (31). It is our belief that considerable advances can be expected in the near future on the understanding and use of electronic and optical properties of thin polymeric films and they will play a major role in the development of optical data processing.

References

1. Gabor, D. *Nature, London* **161**, 777 (1948).
2. Brot, J., and Robillard, J. *C.R. Acad. Sci. Paris,* **T277**, 167 (1973).
3. Jeudy, M., and Robillard, J. *Optics Communications* **13**, 25 (1975).
4. Bergstein, L., and Kermish, D. Volume 1 Holography, *Modern Optics,* Brooklin Polytechnic Institute, New York (1967).
5. Robillard, J. U.S. Patent No. 3,989,530 (1976).
6. Robillard, J. French Patent No. 70,37,371.
7. Berthier, G., and Pullman, A. *C.R. Acad. Sci., Paris* **234**, 152 (1952).

8. Palit, S. *J. Chem. Soc.*, **52,** 459 (1960).

9. Bruhat, G. *Cours de Physique Generale*, T3, Masson, Paris, (1960).

10. Weimann, J. *Relations entre la Structure et les Proprietes Physiques,* Masson, Paris (1965).

11. White, M. *Organic Photoconductors,* J. Wiley, N.Y. (1972).

12. Sharp, J. H., and Smith, M. Organic semiconductors. In *Physical Chemistry.* Vol. 15, Academic Press, New York, (1975).

13. Vartanyan, A. *Acta Physicochim. USSR,* **22,** 201 (1947).

14. Geacintov, N. E., and Pope, M. *Intrinsic Photoconductivity in Organic Crystals.* McGraw Hill, New York (1971).

15. Pankove, J. I. *Optical Processes in Semiconductors,* Prentice Hall, New York (1971).

16. Born and Wolf, *Principle of Optics,* Pergamon Press, New York (1975).

17. Minkin, V., et al. *Dipole Moments in Organic Chemistry,* Plenum Press, New York (1970).

18. Oster, G. K., and Oster, G. U.S. Patent 3,097,097 (1963), Canadian Patent 664,566 (1963).

19. Oster, G. K., and Oster, G. *J. Am. Chem. Soc.* **81,** 5543 (1959).

20. Ya Gel'Fman, Vysomole. *Soyed* **5,** 10, 1534 (1963).

21. Oster, G. K., and Oster, G., *J. Amer. Soc.* **8,** 5543 (1959).

22. Garito, A., and Singer, K. D. *Laser Focus* **80,** 59 (1982).

23. Sauteret, C. et al. *Phys. Rev. Letters* **36,** 956 (1976).

24. Winsor, M. W., et al. *Spectrochim. Acta* **18,** 1364 (1962).

25. Smets, G., et al. *Tetrahedron* **25,** 30251 (1969).

26. Sergio Calixto. Symposium on Industrial Uses of Holography, Las Cruces, New Mexico (1987).

27. Tomlinson, J., et al. *Appl. Phys. Letters* **16,** 486–489 (1970).

28. Hugnard, J. P., et al. *Revue Technique Thomson CSF,* **8,** 4 (1976).

29. Sohn, J. E., et al. *Makrom Chemie* **180,** 2975 (1979).

30. Lalama, S. J., et al. *Appl. Phys. Letters* **39,** 940 (1981).

31. Robillard, J. J. U.S. Patent Application No. 815810 (1986).

V
Recording equipment

12
Holographic optical head for optical disk memories

YUZO ONO, YASUO KIMURA, and SEIJIN SUGAMA

An optical head for optical disk systems, such as write-once optical disks, magneto-optical (MO) disks, compact disks (CD), and laser video disks (VD), has three optical functions, in addition to light beam focusing on a disk. These are splitting the reflected beam, which carries the readout signals from the optical axis, focusing error detection optics, and tracking error detection optics. Conventional optical heads have employed several bulky polished optical elements to realize these three optical functions; for example, a beam splitter prism, a cylindrical lens and a spherical lens pair, and a grating, respectively. This complicated optics set-up has prevented realizing small size and lightweight optical heads.

In optical disk systems, however, lightweight small size optical heads are required for high speed track seeking. For this purpose, an optical head using a holographic optical element (HOE) has been developed (1–3). This HOE combines the mentioned three optical functions. This optical head is able to operate independently from wavelength variation in a light source, despite using the HOE.

This chapter describes principles of error signal detection and beam splitting for holographic optical heads using HOEs, fabrication method for HOEs, and applications for CD, VD, and MO optical heads.

Error signal detection principle

Focusing error signal detection principle

The optical configuration of this head for CD players is shown in Figure 12.1 as an example to explain the error signal detection principle. This optical head consists of only four optical elements. A photo detector (PD) having six sensor segments, S1–S6, is set closely beside the laser diode (LD) used for a light source. The central four segments, S1–S4, are divided by the two division lines parallel to the disk radial direction (r-direction), namely, the X axis, and the disk track direction (t-direction), namely, the Y axis.

The holographic optical element consists of four different holograms. Holograms H1 and H2 have a boundary line parallel to the r-direction. Small holograms H3 and H4 are formed on this boundary line on both sides of the optical axis.

To generate these holograms, four points, P1–P4, are assumed. Point P1 lies on the X axis, which is a boundary between S1 and S3, whereas point P2 lies on the X axis, which is a boundary between S2 and S4. Points P3 and P4 lie on S5 and S6, respectively, as shown in Figure 12.1. In the individual hologram, H1–H4 in Figure 12.1, the interference fringe, generated by a divergent spherical wave, which diverges from the LD, and a divergent spherical wave, which diverges from point P_n ($n = 1$–4), are recorded, respectively.

The operating principle is as follows. A small part of the light emitted from the LD is diffracted out of the optical axis by the hologram, but the light transmitted through the hologram is focused on the disk by the imaging lens. The light reflected on the disk carrying the recorded data is transformed into a spherical wave, converging to the light emitting point on the LD by the imaging lens, and illuminates the holograms. When the light beam is exactly focused on the disk, the incident light on the hologram, H1–H4, is diffracted so as to focus on P1–P4, respectively, as shown in Figure 12.1.

The CD readout signal is obtained by the sum of the S1–S4 output powers.

In this optical head, the focusing error signal is derived by using the principle of double-knife-edge method. The boundary line between holo-

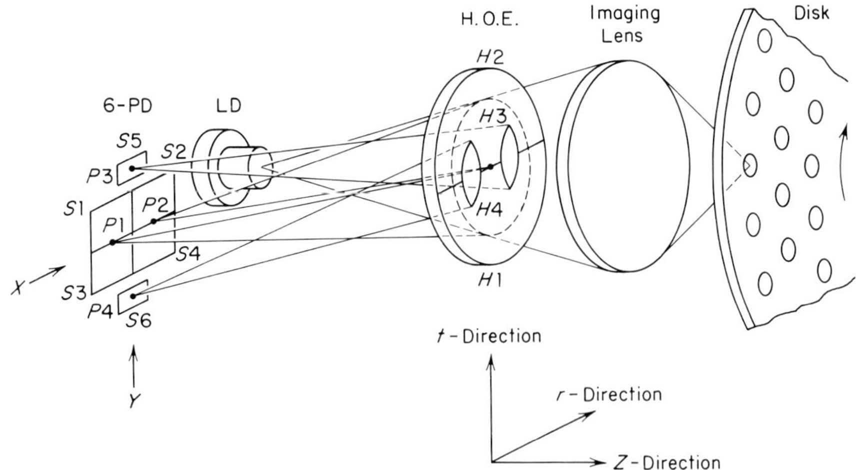

Figure 12.1. Optical configuration of holographic optical head for CD players.

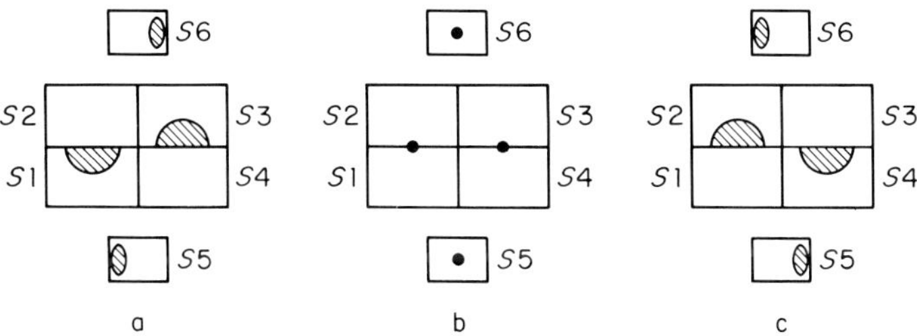

Figure 12.2. Spot shapes variation on PD segments S1–S6 caused by defocusing, (a) when the disk is too close, (b) in focus position, (c), too far away.

grams H1 and H2 acts effectively as a knife edge. Figure 12.2 shows the spot shapes on the PD. When the disk is in focus (Fig. 12.2b), the central two light spots focus in the X axis. When the disk moves too close to the head (Fig. 12.2a), the central two defocused light spots lie on only S2 and S3. On the contrary, when the disk moves too far away (Fig. 12.2c), the two defocused light spots lie on only S1 and S4. Therefore, focusing error signal S_{fe} is given by $S_{fe} = (V1 + V4) - (V2 + V3)$, where V1–V4 are output powers for S1–S4, respectively. As the boundary line for holograms H1 and H2 is parallel to the r-direction, that is, the tracking direction, the tracking error has no influence on the focusing error signal. Therefore, cross-talk between the focusing error signal and tracking error signal is small.

Reduction in wavelength fluctuation influence

Grating optical elements, such as holograms, have not been applied to an optical head because of the sensitivity to wavelength fluctuation in the laser diode used for the light source. In the optical configuration for holograms in this optical head, lateral aberration, that is, variation in diffraction angle due to the wavelength variation in the light source, arises along the r-direction, namely, the X axis, on the PD. The size of each sensor segment for a PD is sufficiently larger than movement length of the focused spots on PD corresponding to wavelength variation. Therefore, the focusing error signal is not affected. The longitudinal aberration, that is, the variation in the hologram focal power, due to the wavelength variation is removed as the holograms are generated by using interference between two almost identical divergent spherical waves to make the focal power be zero.

The wavelength variation influence has been numerically analyzed using the ray-tracing method (4). Figure 12.3 shows the optical configuration and coordinate system. The hologram is on the $X–Y$ plane. The light emitting point on the LD and the light beam converging point on the PD are arranged at $(0, 0, Z_{\mathrm{LD}})$ and $(0, Y_f, Z_f)$, respectively. In the numerical analysis, the following values are used; distance between LD and hologram is 18 mm, distance between LD and the PD center is 7.6 mm, longitudinal

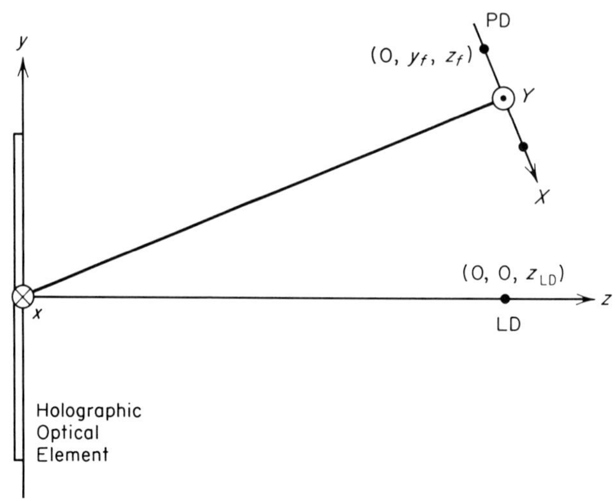

Figure 12.3. Coordinates for calculating an interference fringe pattern.

magnification for the imaging lens is 20.67, and designed wavelength is 780 nm.

The numerical result show that the focusing error signal off-set, due to LD production lot difference, can be removed by the PD position adjustment. The focusing error signal off-set, due to the wavelength during the head operation, was also analyzed. When the wavelength is fluctuated from 780 to 785 nm by 5 nm, the focusing error signal off-set $\Delta F/\Delta\lambda$ is 0.2 μm/5 nm. This value is within an acceptable focusing error tolerance for the optical disk operation. This analysis was carried out for the optical configuration, where $Z_{LD} = Z_f$. The PD, however, is located at a point where $Z_{LD}^2 = Y_f^2 + Z_f^2$, the focusing error signal off-set due to wavelength fluctuation becomes still smaller and is nearly zero.

Tracking error signal detection principle

Tracking error signal S_{te} is derived by using the push–pull method. The incident light power difference between holograms H3 and H4 is detected by the output difference between segments S5 and S6. Therefore, S_{te} is given by $S_{te} = V5 - V6$, where V5 and V6 are output power values for S5 and S6, respectively. A usual push–pull tracking method, however, has signal off-set problems, as shown in Figure 12.4. With any movement of the tracking actuator, as shown in Figure 12.4a, there arises an optical axis discrepancy between the imaging optics and the error detection optics. In such a case, the cross section of the reflected beam on PD1 and PD2 becomes different. Consequently, a signal off-set in the tracking error signal is caused. Any appreciable inclination of the disk will also cause this same type of signal

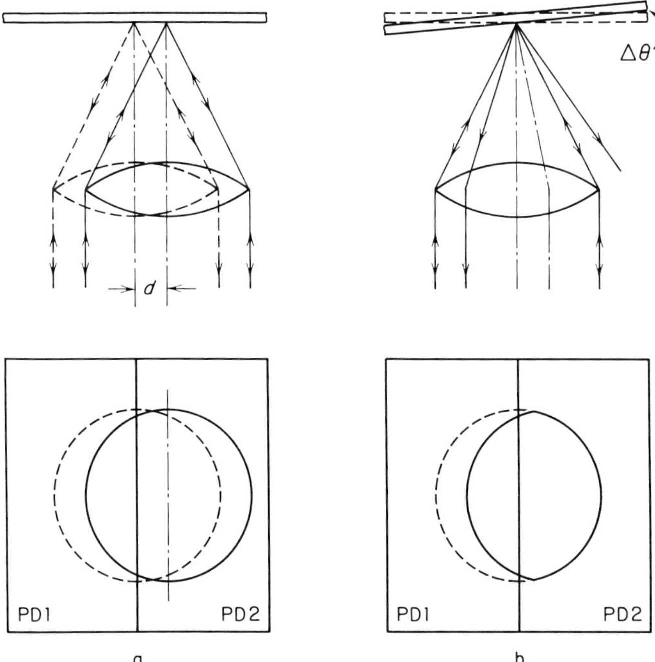

Figure 12.4. Origin of a signal off-set on the tracking error signal. a. Optical axes discrepancy between imaging optics and error detection optics. b. Inclination of the disk.

off-set, as shown in Figure 12.4b. In this optical head, to suppress any off-set on the tracking error signal, holograms H3 and H4 for tracking error detection are located on both off-axis sides on the HOE.

In this configuration, it is possible to greatly reduce the tracking signal off-set by optimizing the shapes of tracking holograms H3 and H4 (2). Figure 12.5 shows the relation between the axis for imaging optics and error detection optics. Tracking holograms H3 and H4 are arranged so that only the most sensitive part for tracking error in the reflected beam (dotted part in Figure 12.5) illuminates H3 and H4, in spite of any tracking actuator movement. The light power distribution for each dotted part of the reflected beam is almost uniform in each dotted part, as a result there is no change in the incident light power for an individual tracking hologram with any tracking actuator movement. Therefore, signal off-set is suppressed. Figure 12.6 shows calculated tracking off-set results both for a conventional push-pull method and for this method. At the calculation, a divergent wave from a usual laser diode, which has Gaussian light power distribution, is assumed. In Figure 12.6, the vertical axis shows the magnitude of the tracking error off-set, defined as the discrepancy between the center level of

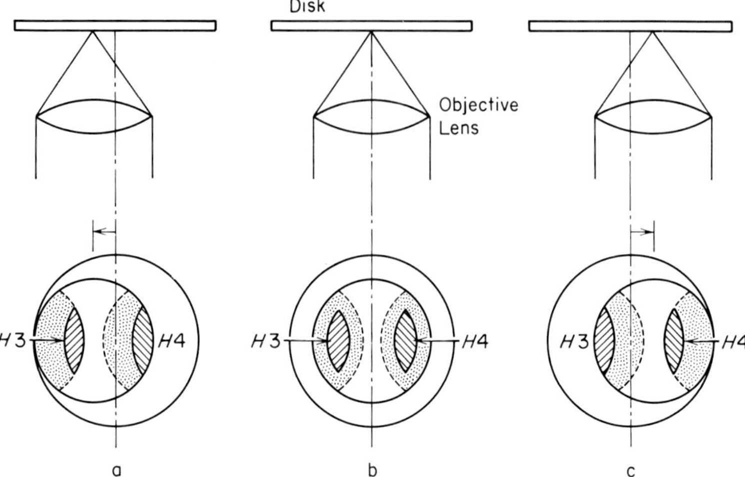

Figure 12.5. Relation between the axes of imaging optics and error detection optics, (a) when the objective lens is moved to the left, (b) in central position, (c) moved to the right.

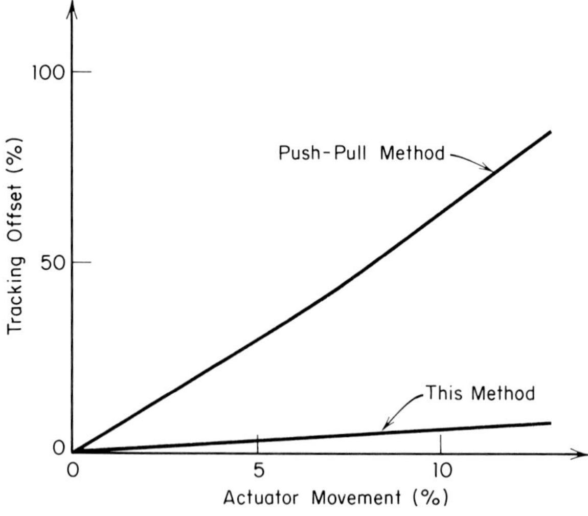

Figure 12.6. Calculated tracking error off-set results for both push–pull method and this method.

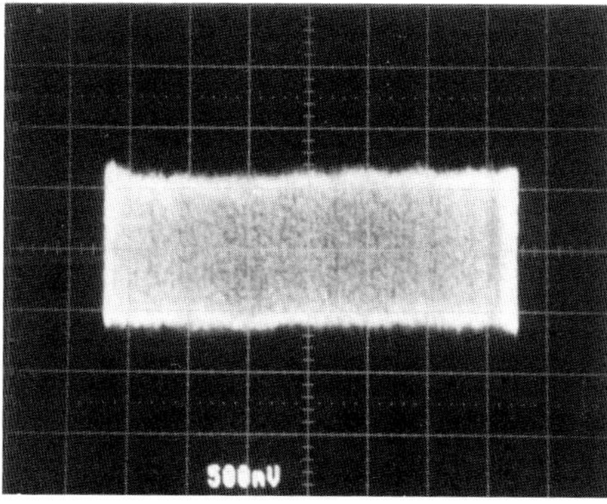

Figure 12.7. Envelope variation in tracking error signal for this method, generated by ±0.6 mm tracking actuator movement.

the tracking error signal amplitude and the ground level, which is normalized by its own amplitude. The horizontal axis shows the tracking actuator movement normalized by the aperture of the imaging lens. Figure 12.7 shows the envelope variation in tracking error signal for this optical head, generated by the 0.6-mm tracking actuator movement. In Figure 12.7, the vertical and horizontal axes show the tracking error signal amplitude and tracking actuator movement, respectively. From these figures, it is clear that this method is very effective to suppress tracking error off-set

Beam splitter function

Beam splitter for compact disk head

In the optical head for CD players, shown in Figure 12.1, both transmitted light and first-order diffracted light by HOE are used. Therefore, the total light power efficiency is given by multiplying the transmission and first-order diffraction efficiencies.

In the case of thin surface relief holograms, which have a rectangular relief profile, the theoretical maximum value of total light power efficiency is $1/\pi^2$ (equal to 10.1 percent). This value is limited, because the efficiency of the minus first-order diffraction light, which is not used in this optical head, has the same value as the efficiency of the plus first-order diffraction light, which is used for signal detection. Therefore, to obtain the higher total light power efficiency, only the plus first-order diffraction light has to be enhanced under the same transmission efficiency. This can be achieved by using holograms that have an asymmetrical relief profile.

When the holograms have a sawtooth relief profile, the total light power efficiency has a theoretical maximum value, which is 16.2 percent (5). The

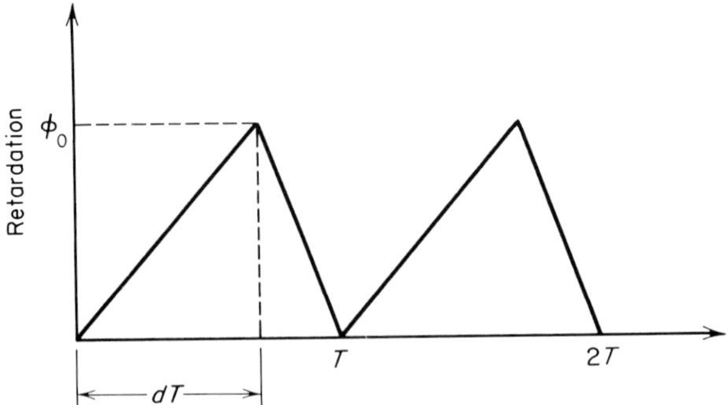

Figure 12.8. Relief profile for actual holograms.

relief profile shape for an actual hologram has some deviation from the sawtooth relief profile, as shown in Figure 12.8. In this figure, d shows the degree of this deviation. The theoretical relation between d and total light power efficiency η is shown in Figure 12.9. In this figure, phase retardation between top and bottom of the hologram ϕ is taken for a parameter. In the hologram fabricated by oblique incident angle ion beam etching (IBE)

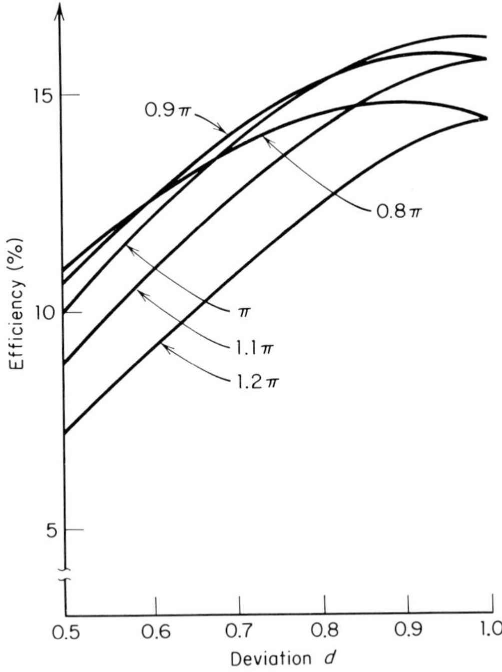

Figure 12.9. Relation between efficiency and deviation d.

method (6), d was about 0.7 to 0.9. Under this condition, high efficiency, which exceeds 14 percent, can be realized when the retardation is about $0.8\pi - \pi$. Therefore, fabrication conditions are not so severe.

Polarizing beam splitter for magneto-optical disk head

The polarizing beam splitter (PBS) is an indispensable optical element for MO heads. A compact MO head can be realized by adding the PBS function to the HOE used for a CD head. To apply the HOE for the MO head, the HOE must deliver the polarized light component that contains the Kerr signal to the RF signal detecting optics without appreciable light power loss. This means that only the polarized light component, which is perpendicular to the mentioned polarization, is delivered to the error detection optics. Some fraction of light power must be delivered to the error detection optics, because the disk sector index signal, which is recorded on the disk in pit depth form, is detected from the error detection signals in usual MO heads. When the Kerr signal component is assumed to be s-polarization, HOE must have PBS characteristics shown in Figure 12.10.

Because the high spatial frequency diffraction grating, whose pitch d is smaller than wavelength λ, does not generate higher-order diffracted light, it is suitable for use as a beam splitter. The diffraction efficiency for such gratings shows large polarization dependence (7). A grating with $\lambda/d > 1.6$ has nearly zero diffraction efficiency for TM wave (polarization perpendicular to the grating grooves), but the diffraction efficiency for TE wave (polarization parallel to the grating grooves) is varied with differing grating groove depth h value. For example, when $\lambda/d = 1.6$ and $h/d = 0.5$ for Bragg diffraction geometry, diffraction efficiencies are $\eta_1(\text{TM}) = 0$ percent, $\eta_1(\text{TE}) = 20$ percent, $\eta_0(\text{TM}) = 100$ percent, and $\eta_0(\text{TE}) = 80$ percent. This distribution just matches the required PBS characteristics in the MO head shown in Figure 12.10.

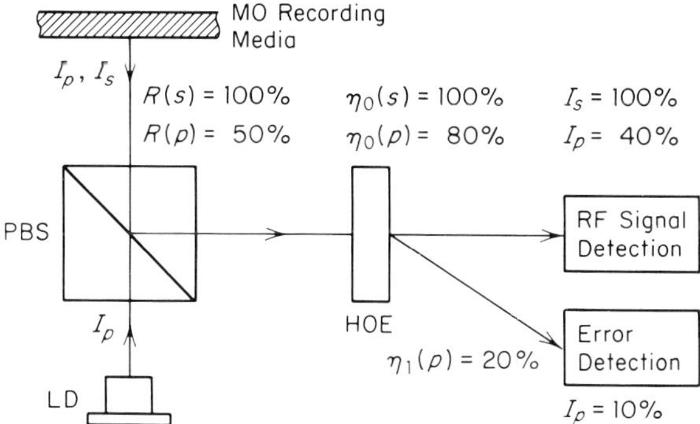

Figure 12.10. Polarizing characteristics required for HOE used for MO head.

Holograms fabrication process

The original drawing is fabricated, using electron beam (EB) lithography, to arrange the four holograms precisely. The interference pattern for the holograms drawn by EB is a hyperbolic group, generated by the interference between two divergent spherical waves. In the coordinates shown in Figure 12.3, the equation for the interference fringe pattern is

$$\frac{\left(Y+\dfrac{T}{2Q}\right)^2}{\dfrac{1}{Q}\left(\dfrac{T^2}{4Q}-U\right)} - \frac{X^2}{\dfrac{1}{4P}\left(\dfrac{T^2}{4Q}-U\right)} = 1 \tag{12.1}$$

where P, Q, T, and U are constants determined by the arrangements of two

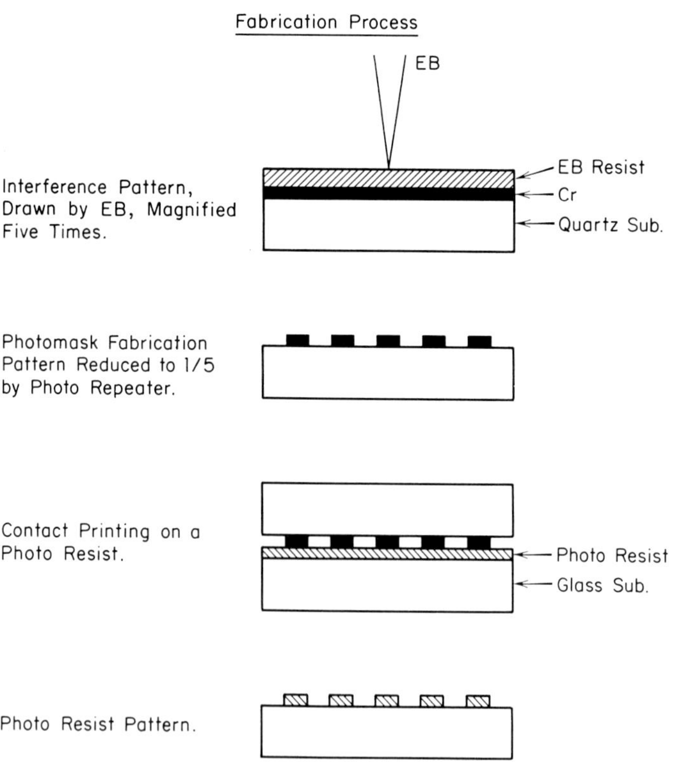

Figure 12.11. Holographic optical element fabrication process.

light emitting points and the wavelength, which are given by

$$P = (\lambda n)^2$$

$$Q = 4Y^2_{PD} - 4(\lambda n)^2$$

$$T = 4Y^2_{PD}\{Z^2_{LD} - (Y^2_{PD} + Z^2_{PD}) + (\lambda n)^2\}$$

$$U = Y^2_{PD} + Z^2_{PD} + Z^2_{LD} + (\lambda n)^2 - 4(\lambda n)^2 Z^2_{LD}$$

$$(12.2)$$

where, λ is the wavelength of the LD and n is an integer.

The fabrication process is shown in Figure 12.11. First, the pattern, magnified five times, is drawn by EB. A photomask on which many holographic optical elements are arranged, is fabricated by reducing the magnified pattern to one fifth using a photorepeater. This reduction process is effective to reduce stage error and the quantization error in the EB drawing. Holographic optical elements are easily duplicated by contact printing in a photoresist layer on the glass substrate, using this photomask. The fabricated hologram is 6 mm in diameter. Holograms H3 and H4 are

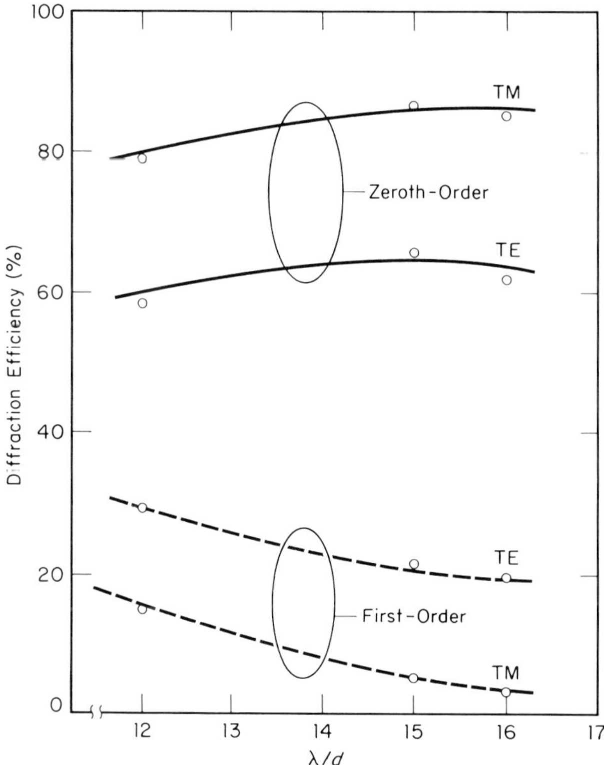

Figure 12.12. Polarization dependence of diffraction efficiency versus grating pitch.

formed 0.75 mm from the optical axis. An individual area occupies 7.2 percent in the cross section of incident reflected beam from the disk. Average pitch for the hologram is about 2.5 μm. For the hologram etched using the IBE method, the total light power efficiency was 14.8 percent. This value includes reflection loss.

To fabricate the polarizing HOE for MO heads, the original photomask is fabricated using direct EB drawing. In the practical fabrication process, to correct any astigmatism caused by an oblique incident angle substrate, the astigmatic wave is assumed instead of a spherical wave. The interference fringe pattern, therefore, is a complicated implicit function. The averaged pitch for the interference fringe is designed to be 0.52 μm, so that $\lambda/d = 1.6$ at $\lambda = 0.83$ μm. Holographic optical elements are duplicated in a photoresist layer on the glass substrate using this mask. The fabricated hologram has a 6×10-mm dimension ellipse shape.

Figure 12.12 shows the diffraction efficiency distribution for the fabricated HOE, whose groove depth $h = 0.25$ μm. At the designed wavelength ($\lambda = 0.83$ μm, $\lambda/d = 1.6$), the polarizing characteristics required for the MO head are obtained.

Applications for optical disk heads

Application for compact disk head

The optical configuration and operating principle have already been explained, using Figure 12.1. The fabricated holographic optical head is shown in Figure 12.13. The focusing error detection characteristics are shown in Figure 12.14. The linear signal detection region for the focusing error detection characteristics is about 20 μm.

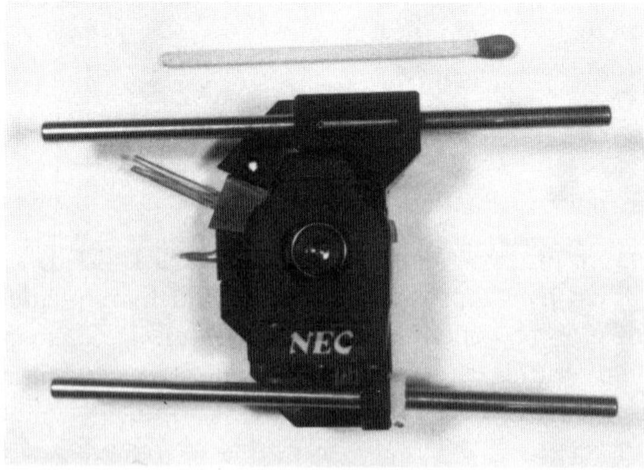

Figure 12.13. External view of fabricated holographic optical head for CD players.

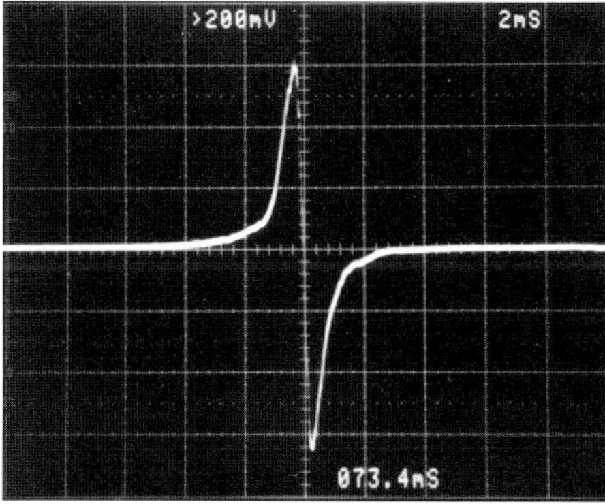

Figure 12.14. Focusing error detection characteristics.

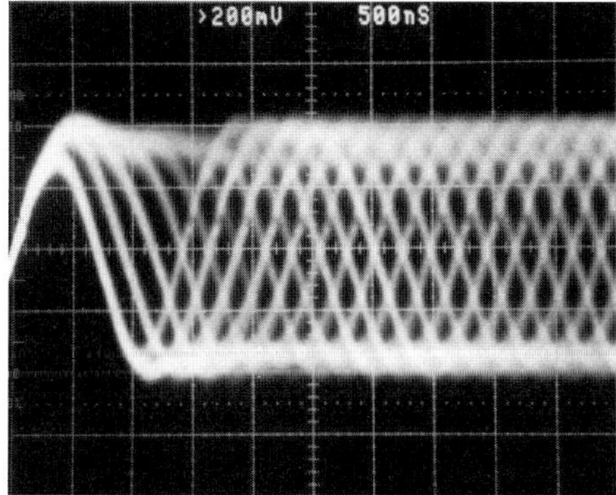

Figure 12.15. CD readout signal.

Signal play back from the CD was realized with this optical head using a tracking and focusing error servo. The readout signal from this optical head typically has about 18 nsec jitter for 3T signal. The readout signal from a CD is shown in Figure 12.15.

Application for laser video disk head

The optical isolator function is required for a VD head to attain a readout with high S/N ratio by suppressing the reflected light incident on the LD, because the signal is recorded in analogue form. In this optical head, an optical isolator consisting of a PBS and a quarter wave plate, is used. The HOE is used in the light split by the PBS. The optical configuration for the head is shown in Figure 12.16. The HOE acts as a beam splitter and the RF readout signal is derived from the zeroth-order diffracted light. The focusing

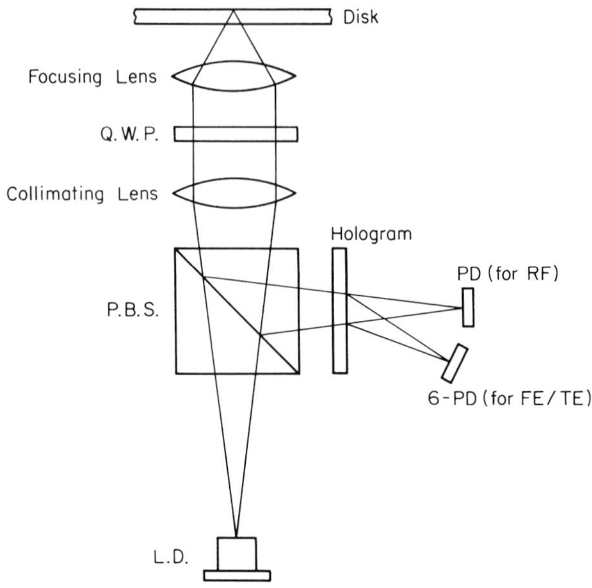

Figure 12.16. Optical configuration of holographic optical head for laser video disk players.

and tracking error signals are derived from the first-order diffracted light. For the light source, a visible light LD, composed of AlGaInP, whose wavelength range was 660 to 680 nm, was used (8). The HOE was designed at wavelength $\lambda = 670$ nm. The S/N ratio for the readout signal measured by the video S/N meter was about 42 dB. Figure 12.17 shows the readout signal spectrum.

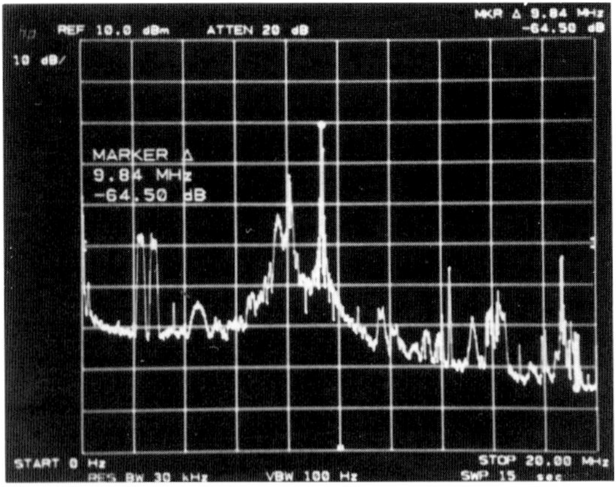

Figure 12.17. Readout signal spectrum for VD.

Application for magneto-optical disk optical head

The optical configuration for an MO head, using the polarizing HOE, is shown in Figure 12.18. The light reflected back onto the lens from the disk is split by the PBS prism. Its polarization direction is rotated by the half-wave plate, so that the Kerr readout polarization component becomes TM polarized with regard to the HOE. Then, it becomes incident on the HOE at the Bragg angle. The first-order diffracted light, which is split into four beams, is used for error signal detection. The zeroth-order diffracted light is used for signal readout through the polarizing prism.

The HOE consists of four different holograms, the same as that for CD and VD heads. Focusing and tracking error signals are derived using these principles. Kerr readout signal is derived from the two-segment photodetector using a differential method.

In read and write experiments using the proposed MO head with the HOE, more than 57 dB C/N ratio was obtained for pit width recording at 1 MHz recording frequency, as shown in Figure 12.19.

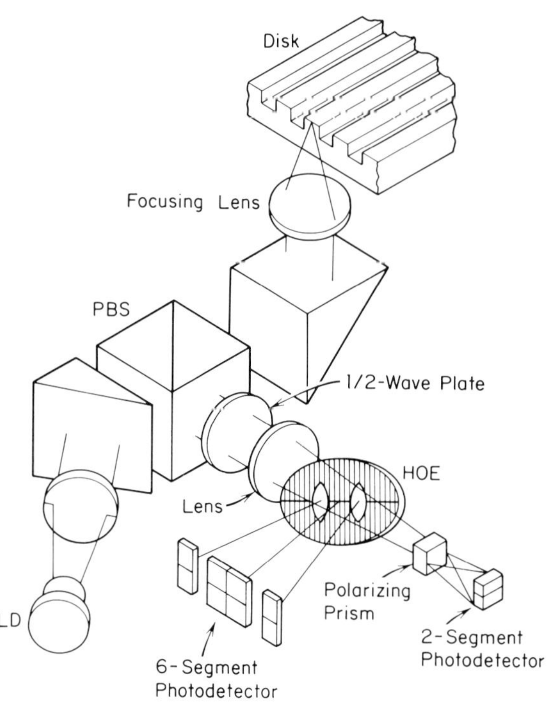

Figure 12.18. Optical configuration for MO head using a polarizing HOE.

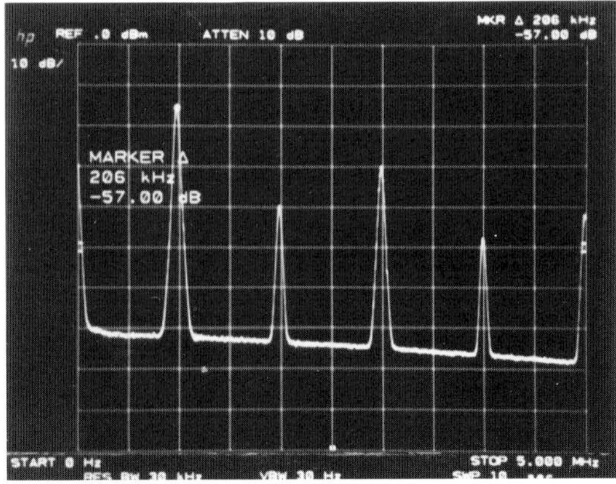

Figure 12.19. Readout signal spectrum for MO disk.

Conclusion

A new concept optical head for optical disks has been developed using the holographic optical element. The holographic optical element has three optical functions required for an optical head: splitting reflected beam, focusing error detection, and tracking error detection. In addition to these, a polarizing beam splitter function, using a high spatial-frequency hologram, is also combined for a MO disk head. This optical head is free from wavelength variation, in spite of the hologram use. The off-set signal in the tracking error signal has been reduced, using an optimized holographic optical element. In applications for CD, VD, and MO disks, stable operations with high readout S/N ratio have been obtained.

References

1. Kimura, Y., Sugama, S., and Ono, Y. Tech. Digest of 4th Topical Meeting on Optical Data Storage, 178 (1987).
2. Kimura, Y., Sugama, S., and Ono, Y. Int. Symp. on Optical Memory, Tokyo, September (1987).
3. Kimura, Y., and Ono, Y. First Micro Optics Conference, Tokyo, October (1987).
4. Holloway, W. H., and Ferrente, R. A. *Appl. Opt.* **20,** 2081 (1981).
5. Magnusson, R., and Gaylord, T. K. *J. Opt. Soc. Am.* **68,** 806 (1978).
6. Kosuge, K., Sugama, S., Ono, Y., and Nishida, N. Conference Digest of 13th ICO, 526 (1984).
7. Yokomori, K. *Appl. Opt.* **23,** 2303 (1984).
8. Kobayashi, K., Kawata, S., Gomyo, A., Hino, I., and Suzuki, T. *Electron. Lett.* **21,** 931 (1985).

VI
Teaching of holography

13
The teaching of holography to industry-oriented students

VINCENT TOAL

This chapter discusses the experience we have had over a 7-year period in the training of fourth year undergraduate students in holography and related skills. Project work plays an important role here and a number of such projects are outlined. The results of some of the more successful are summarized.

Structure of the courses

Context

The holography lectures form part of a lecture course of 30 hours duration in Applied Optics that takes place in the final year of a 4-year degree in Applied Sciences. The degree course is designed to equip students with a range of technological, management, and language skills. It is founded on the conviction that in a rapidly changing industrial and economic environment, a range of skills is essential. In addition, each of the major options (physics, mathematics, chemistry) is studied in depth to provide adequate preparation for those wishing to specialize or undertake graduate research. Table 13.1 outline the overall degree course structure.

Table 13.1. Applied sciences degree course structure

Year 1	Physics Chemistry Mathematics Language Industrial Studies
Year 2	Two from: Physics Chemistry Mathematics plus Language and Industrial Studies Students taking Physics and Chemistry also take ancillary Mathematics
Year 3	As in Year 2
Year 4	Two from Physics, Chemistry, and Mathematics

Students major in two of three subjects, physics, chemistry, or mathematics. Within these, optional specialized topics are offered in the final year including optoelectronics. The fourth year physics course structure is outlined in Table 13.2.

Table 13.2. Physics course year 4

Solid State Physics
Statistical Physics
Electrical and Electronic Instrumentation
Applied Optics

Options (two from)
    Radiation and Nuclear Physics
    Acoustics
    Lasers and Optoelectronics
    Topics from Applied Biophysics

Each course, 30 lectures. Laboratory 7–14 hours weekly

Typical first-year student intake now stands at about 80 and the final year class in 1986–1987 comprised 65 students of whom 26 took the physics and mathematics option and 22 the physics and chemistry option.

Applied optics course structure

The course is demanding and account must be taken of the fact that only half the student body are concurrently taking mathematics. For this reason the mathematical tools are as far as possible integrated into the course. All students will have been taught Fourier analysis.

The course is outlined in Table 13.3.

The technique of Fourier transform spectroscopy is dealt with in some detail, including problems associated with the Nyquist sampling criterion, finite signal duration, and wavelength resolution. A structured laboratory experiment is provided that uses various light sources, a Michelson interferometer, and photodetector/amplifier. Signals are passed to a storage oscilloscope interfaced with an Acorn BBC 32K microcomputer with dual floppy disk drives and a dot matrix printer. This machine has proved very versatile in its use in the undergraduate laboratory. A comprehensive Fourier software package is also provided including graphics facilities. In this way students are encouraged to use Fourier methods in optics.

Further lectures on Fabry–Perot interferometry and thin multilayer films are given before diffraction and the concept of spatial frequency are introduced. Complex notation is used.

Various diffracting systems are considered in detail using the Fraunhofer diffraction formula. The mathematical equivalence of the Fraunhofer formula and the two-dimensional spatial Fourier transform is demonstrated.

Table 13.3. *Applied optics*

Interferometry

Fourier transform spectroscopy. Multiple beam interferometry. Fabry-Perot etalon and its uses, scanning Fabry-Perot. Matrix theory of thin films. Antireflection coatings, narrowband transmission filters, and multilayer dielectric mirrors.

Diffraction

Theory of diffraction gratings. Grating spectroscopy. Echelons. Application of grating equation to antenna design. Circular apertures, lenses, resolution, and apodization.

Fourier Optics

The equivalence of Fraunhofer diffraction and 2-D spatial Fourier transformation. Fourier transformation by a lens. Spatial frequency. Applications of spatial filtering, such as scan line removal, focusing error correction, detection of flaws in integrated circuit masks, phase contrast and dark ground microscopy, schlieren technique, edge enhancement, mathematical operations, and character recognition.

Holography

Historical overview. Basic theory of holographic recording and reconstruction. The photographic process.

Requirements relating to the source and recording material. Problems of thermal and mechanical stability. Amplitude and phase holograms. Volume holograms. Processing techniques.

Holographic applications including display, image plane and rainbow holograms, microscopy, nondestructive testing, aberration correction, and imaging through distorting media.

Non-silver recording schemes. Dichromated gelatin, photo-resist, thermoplastics, photo-polymer systems, and photorefractive materials. Applications of the conjugate wave.

Students (and others) have considerable difficulty in coming to grips with the idea of spatial frequency. I believe this may be due to the often long derivations of the Fourier transforming property of a lens given in a number of textbooks, which leads to the problem of not being able to see the wood for the trees. A short and simple derivation is used (1), which emphasizes the very interesting physics involved. At this point the student is able to grasp the meaning of Figure 13.1 in spatial frequency terms.

The process of image formation as a two-stage diffraction or Fourier transformation process is now described. It is emphasized (Fig. 13.2) that the spatial frequency spectrum of the input signal can be altered at will and one can consider a number of practical uses for spatial filtering, including flaw detection (2), focusing error correction (3), phase contrast and dark ground microscopy (4), mathematical operations, and image enhancement (5). Character recognition (6) can be considered now or left until holographic principles have been dealt with.

Students seem to find the applications of spatial filtering one of the most stimulating topics of the course and this does help considerably in overcoming the lack of mathematical skills in weaker students.

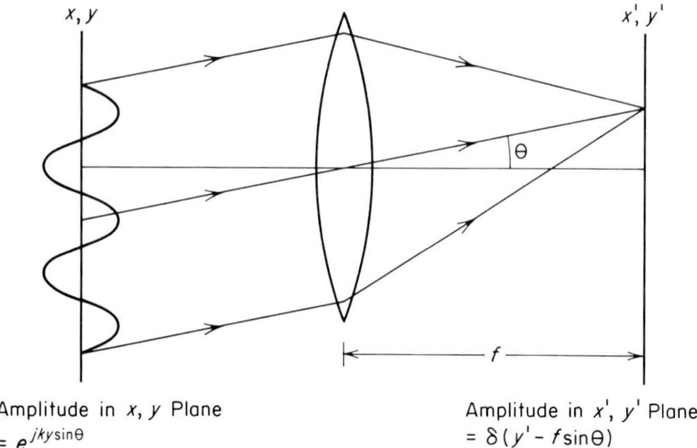

Figure 13.1. Spatial frequency.

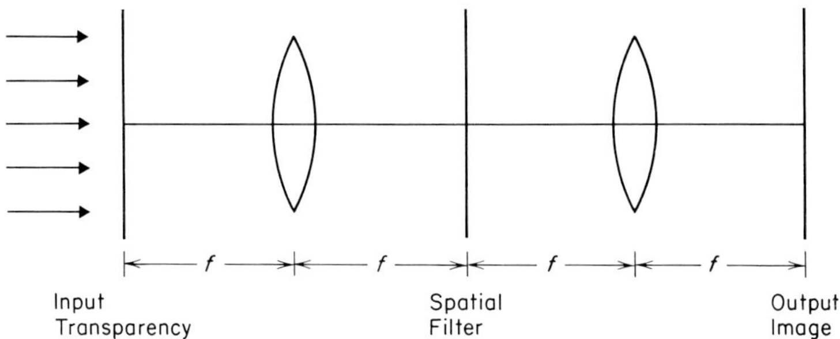

Figure 13.2. Spatial filtering.

Holography

Some historical background is given with emphasis on holography as a two-stage diffraction process. It helps at this stage if the arrangement for recording a hologram is seen as an interferometer of the Mach–Zehnder type with one of the mirrors replaced by the object to be recorded and the output beam splitter replaced by a photographic plate (Fig. 13.3). Indeed all holographic recording systems are interferometers with geometrical variations.

The theory of the photographic process for silver halide is outlined and an expression obtained for the amplitude transparency of the processed emulsion (Fig. 13.4). The disturbance produced in a plane by the interference of a simple wave front (reference wave) and a coherent complicated one (object wave) is calculated (Fig. 13.5a).

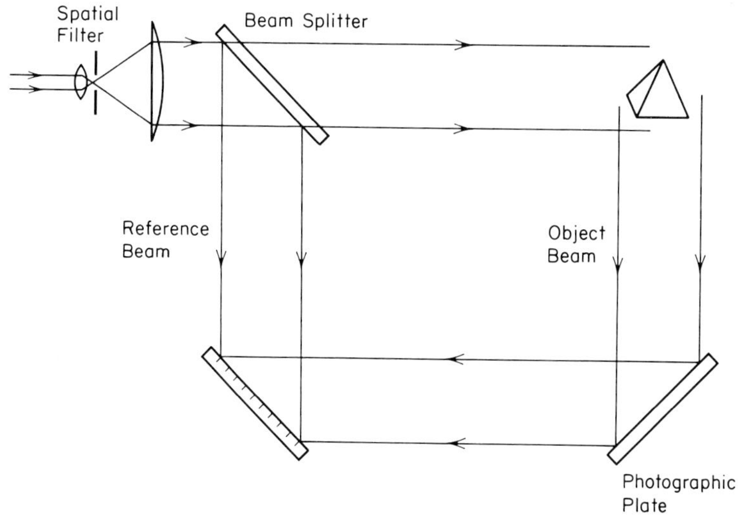

Figure 13.3. Mach–Zehnder holographic arrangement.

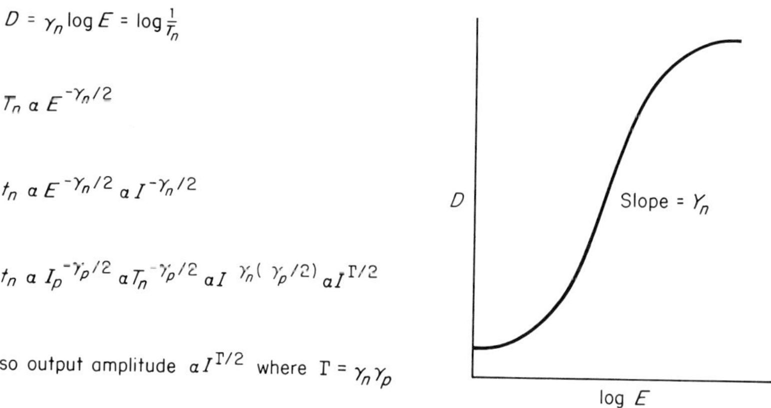

$$D = \gamma_n \log E = \log \frac{1}{T_n}$$

$$T_n \ \alpha \ E^{-\gamma_n/2}$$

$$t_n \ \alpha \ E^{-\gamma_n/2} \ \alpha \ I^{-\gamma_n/2}$$

$$t_n \ \alpha \ I_p^{-\gamma_p/2} \ \alpha \ T_n^{-\gamma_p/2} \ \alpha \ I^{\gamma_n(\gamma_p/2)} \ \alpha \ I^{\Gamma/2}$$

so output amplitude $\alpha I^{\Gamma/2}$ where $\Gamma = \gamma_n \gamma_p$

Figure 13.4. Photographic process.

One can now demonstrate the existence of zero order, reconstructed object wave, and conjugate wave, by combining the results of this discussion (Fig. 13.5b). The advantages of off-set over in-line reference beam arrangements are considered, particularly the spatial separation of the three beams.

Practical matters, such as the necessity for a high degree of coherence in the illuminating source, resolution capability of the recording material, and the need for thermal and mechanical stability, are explained. A brief survey

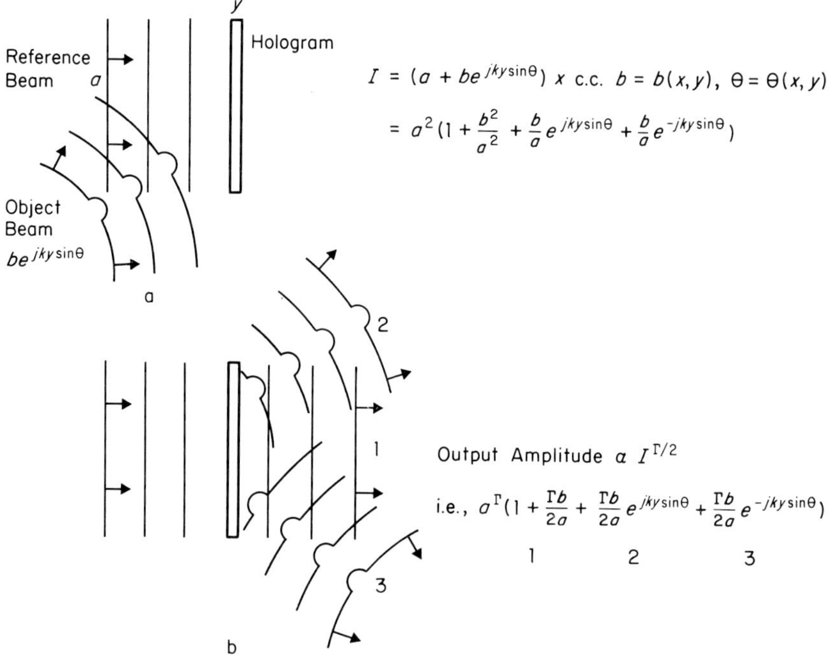

$$I = (a + be^{jky\sin\theta}) \times \text{c.c.} \quad b = b(x,y), \quad \theta = \theta(x,y)$$

$$= a^2\left(1 + \frac{b^2}{a^2} + \frac{b}{a}e^{jky\sin\theta} + \frac{b}{a}e^{-jky\sin\theta}\right)$$

Output Amplitude $\alpha\ I^{\Gamma/2}$

i.e., $a^{\Gamma}\left(1 + \dfrac{\Gamma b}{2a} + \dfrac{\Gamma b}{2a}e^{jky\sin\theta} + \dfrac{\Gamma b}{2a}e^{-jky\sin\theta}\right)$

$$\qquad\qquad 1 \qquad\qquad 2 \qquad\qquad 3$$

Figure 13.5. Hologram recording in silver halide layer.

of available laser systems and their characteristics as well as photographic emulsions is given.

Hologram types and their applications are discussed in some detail beginning with Fourier holograms including lensless Fourier holograms. Particular emphasis is given to display holography as most students are required to make this type of hologram in the laboratory. Among other applications considered are holographic microscopy for the study of aerosols and pollutants (7), holographic interferometry including live and frozen fringe techniques, and time-averaged holographic interferometry (8).

There is some difficulty in understanding the meaning of phase conjugacy even among graduate students. The topic is considered in some detail at this point. One first examines the making of a hologram of a point source of light using an off-axis collimated reference beam (Fig. 13.6). The concept of phase lag and phase lead needs to be reemphasized. It can be shown that when a collimated reference beam traveling in the reverse direction is used to reconstruct the hologram, then a wave is produced that must converge toward the original object source point. When one considers a collection of such object points the meaning of the term *pseudoscopic image* becomes clearer. It is also possible to examine the effect of simply inverting the hologram twice in both the collimated and uncollimated reference beam cases.

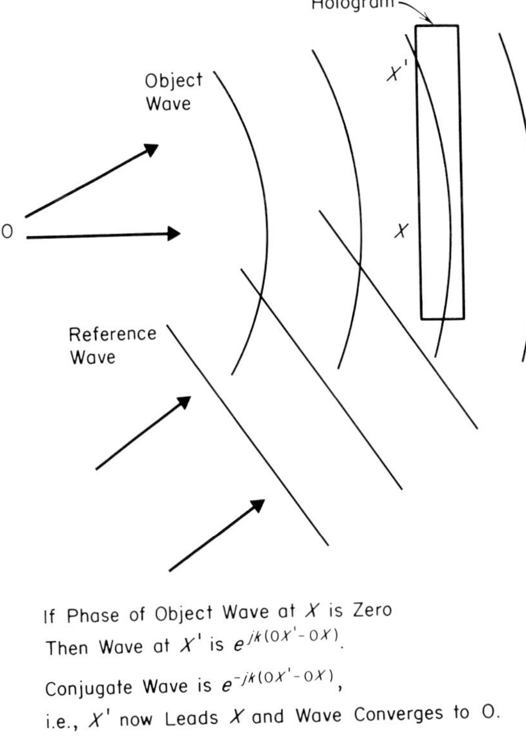

Figure 13.6. Phase conjugacy.

The applications of conjugate image holography are discussed, including image plane and Benton (rainbow) holograms (10), imaging through distorting media (11), and lens aberration correction (12). Real-time applications are described following a survey of other recording media as alternatives to silver halide including dichromated gelatin, photoresists, photopolymer systems, thermoplastics, and photorefractive devices.

Laboratory facilities

Equipment

Our holography laboratory is located on a concrete basement floor of the college building, which in turn is in an area of high density traffic in Dublin city. We have been reluctant to embark on the construction of massive optical supports until the facility is given some security of tenure. The recent construction of an additional college building may help toward this end.

A Gaertner–Jeong metal table with some magnetically clamped optical stands is our mainstay. In addition, there are a few heavy metal stands. The table rests on inflated inner tubes that in turn rest on a solid wooden butcher's block table whose legs are embedded in buckets of sand. Normally our laser, a 10-mW He–Ne (Scientifica-Cook), is mounted on the optical table but for experiments requiring more space the laser is mounted on a separate massive wooden table.

In general students are recommended to carry out exposure of holograms at night when the building is relatively quiet. Thermal currents are the main problem. The laboratory (6.85×1.9 m) has no radiator heating for this reason.

Normally students work singly but in recent years the number of students has risen steeply and pairing has proved necessary. Because the general optics laboratory is operating simultaneously, students are left to their own devices for much of the time and the need for meticulous recording of each step carried out in an experiment is stressed. A series of experiments starting with a single beam transmission hologram and ending with a double object beam reflection hologram are carried out.

To help get started a fairly detailed instruction sheet is supplied. The need for mechanical and thermal stability, however, are demonstrated in the laboratory rather than simply stated. This is easily done by setting up a Michelson or similar interferometer using the laser beam as the source and passing the output beam through a lens to project the fringe pattern on the laboratory wall.

Material costs are rather high and there is considerable wastage. In general film is used rather than plates, the film being supported between glass plates. It is hoped to replace this arrangement by a vacuum platen in the near future. Processing baths are made up by the student using formulations provided.

Total time spent in the holography laboratory ranges from a minimum of 7 hours to a maximum of 20 hours depending on the student's progress. Fifteen hours of laboratory time are scheduled weekly for this class group with additional hours available at the tutor's discretion. Students who wish to carry out additional experiments, for example, in holographic interferometry are welcome to do so if time is available.

Processing

For convenience a developer is used (Table 13.4) that works well for both transmission and reflection holograms.

For transmission holograms any commercial fixer without hardener is used. If bleaching is required, a potassium ferricyanide bleach is used.

For reflection holography a parabenzoquinone bleach is used; the parabenzoquinone being dissolved in acetone to minimize the dangers.

We have used the Pyrochrome process (13) extensively, without using

Table 13.4. Developer Formula

Part A
   100 g ascorbic acid
    50 g sodium sulphite
       dissolve in 5 L deionized water
Part B
   100 g potassium hydroxide
      dissolve in 5 L deionized water

Mix equal parts A and B just before use.

---

color shifting procedures but have found that warm air drying produces a permanent shift in reconstruction wavelength toward the green. While natural drying avoids this problem it is much too time consuming for our purpose. The problem is not nearly so severe with the former process.

To maintain uniform (room) temperature for all solutions and rinses, a 10-L container is filled with deionized water and left in the laboratory the day before a laboratory session.

Other experiments

In addition to the Fourier transform spectroscopy experiment already outlined, a number of other experiments are available in the general optical laboratory and are listed in Table 13.5.

Table 13.5. Optical experiments*

---

1. Speckle metrology
2. Fresnel diffraction
3. Zeeman splitting
4. Fourier transform spectroscopy
5. Scanning Fabry–Perot interferometer
   a. Measurement of free spectral range
   b. Mode structure of a helium–neon laser
6. Refractive index of air
7. Faraday rotation
8. Birefringence

---

*The experiments are representative of the resources available rather than the ideal.

Projects

All students are required to undertake a project that may be in either of their major subjects. Normally project proposals are made by staff but students may submit their own proposals. Additional proposals are made by industrial companies. Projects are assessed with regard to their feasibility, relevance, standard, and the availability of essential equipment.

Following is a list of projects in, or relating to, holography carried out in recent years, with a brief summary in each case.

A Study of the Resolution Attainable in Holography
Direct measurements were made of wire and pinhole holographic images and compared with the object sizes to estimate the line and point spread function width of holograms. Depth of field measurements were also obtained.

Holographic Optical Character Recognition
Fourier transform holograms were recorded of individual characters and a Fourier optical system was set up to search for the presence of the characters in a text. To avoid the problems associated with obtaining precise superposition of the matched filter and the text Fourier transform, a simple apparatus was constructed to enable holographic plates to be processed in situ.

Optical Information Processing
Sinusoidal amplitude transmission filters were made by photographing the fringes obtained from a Lloyd's mirror arrangement. The filters were used to implement an optical addition and subtraction scheme (5). A theoretical analysis of the problem indicates some of the difficulties associated with implementing a practical scheme.

A Schlieren System for the Detection of Surface Defects in Float Glass
This proposal originated with a local optical manufacturing firm. A successful prototype system was developed in which light diffracted by defects, such as scratches, pits, and smears, was converted to an electrical accept/reject signal. Further development funding was subsequently provided by the National Board for Science and Technology.

Speckle Metrology
The technique of speckle photography was used to measure small translations, rotations, and vibrations, both in-plane and out-of-plane.

Image Plane Holography
Collimated reference beam master transmission holograms were made and used to produce image plane white light reflection holograms. A large number of trials were made with various geometrical arrangements using limited facilities.

False Color Holography
The technique of preswelling of the emulsion for the production of color holograms (13) was investigated. Thirty holograms were finally produced using the Denisyuk (14) method. Of these, six showed good quality two-color reconstruction.

Multiple Image Holography
Proposed by the student, this project was aimed at determining the number of plane images that could be successfully stored on a transmission hologram.

Holographic Tomography
The problems associated with the holographic synthesis of three-dimensional images from tomograms were investigated. Holographic reciprocity failure (15, 16) in Agfa-Gevaert emulsions was studied. Some success in producing holograms from a total of four tomograms in each case was achieved. We would like to record our thanks to Dr. Kristina Johnson for valuable discussions.

Holographic Interferometry
Techniques were developed for consistent production of good quality holographic interferograms in an unsuitable environment and without the need for elaborate vibration isolation. Elastic constants of metal samples were measured from the interferograms and compared with mechanical measurements. Figures 13.7 and 13.8 are examples of the interferograms obtained.

The success of this project has enabled students to make transmission holograms using an additional facility in our third year laboratory,

Figure 13.7. Vibration analysis on Interferogram.

Figure 13.8. Vibration analysis on Interferogram.

which is on the first floor and which was not previously found suitable from the point of view of thermal and mechanical stability.

Finally we report recent advances in the technique of silver halide-sensitized gelatin for holographic recording (17, 18) which are the result of a student project carried out in recent months.

Silver halide-sensitized gelatin holography

Silver halide-sensitized gelatin (SHSG) offers an attractive alternative to dichromated gelatin because of its much greater sensitivity, panchromatic response in the case of certain emulsions and greater ease and safety in handling.

The process involves cross-linking of the gelatin molecules and is triggered by the reaction products formed near sites where silver is developed during processing. Hariharan (18) has shown that developer reaction products play a significant role in the formation of the image. In the bleaching step the process of forming a latent image in gelatin is completed.

The emulsion is now treated with an acid fixer without hardener to remove all the silver salts leaving pure gelatin, which is now processed by soaking in alcohol to remove all water. This rapid drying is believed to cause strains in the gelatin ultimately leading to cracking and the appearance of voids in regions of low exposure where the gelatin is not so highly cross-linked and, therefore, weakest (19). There is also evidence (20, 21) that molecular deformation and rearrangement during dehydration produces a density and hence refractive index modulation.

Transmission holography

The 10-mW He–Ne laser was used to make transmission display holograms (22) the object(s) being illuminated by two beams. An exposure of 0.5 to 0.8 J/m^2 was normally given. The processing schedule is given in Table 13.6.

Table 13.6. SHSG processing

1. Develop (Kodak D-19 developer)	2–3 min
2. Rinse (distilled water)	1 min
3. Bleach	until clear plus 1 min
4. Rinse (distilled water)	1 min
5. Fix (acid fixer without hardener)	4 min
6. Wash (running water at 30–35°C)	10 min
7. Rinse (ethanol)	2 min
8. Rinse (propan-2-ol)	2 min
9. Rinse (propan-2-ol)	2 min
10. Dry (dessicator)	3 hours

All solutions at 20°C except at step 6.

The bleach bath consisted of 1.0 g potassium dichromate and 1.0 ml sulphuric acid per litre of distilled water. This schedule follows that of Hariharan (18) except for the wash temperature. It was found that the reduced temperature was necessary for satisfactory results using Agfa-Gevaert 8E75 HD NDH plates and film. This may be due to the fact that this emulsion is rather softer than Kodak 649F.

The holograms were remarkably clean and gave reconstructions that were very sharp and bright. Diffraction efficiency was measured at 0.2 in most cases.

White light reflection holography

In general SHSG holograms have been made having spatial frequencies up to 1000 cycles/mm (18) and 1500 cycles/mm (17). It has been suggested (23) that the removal of silver and silver salts during the fixing process would leave large voids causing increased scatter and that the emulsion would collapse and produce a large shift in reconstruction wavelength. Thus SHSG is not considered to be suitable for white light reflection holography.

Because there remains some doubt about the specific mechanisms involved in gelatin processing, we decided to record holograms using the simple Denisyuk method, retaining the processing method previously described. After a number of trials it was found that exposures of 24 J/m² were optimal.

The reconstructed images were very bright and exhibited little scatter. It was observed that the dominant color of the image changed in a few minutes from red to yellow/gold. Over a period of some days the color changed to green/blue, in an atmosphere of around 70 percent relative humidity. When the holograms were sealed by epoxy resin bonding to a glass plate, the color shifting could be halted. Film holograms were sandwiched between glass plates and sealed with epoxy resin.

Diffraction efficiencies of 0.33 were obtained with the holograms when sealed immediately upon removal from the dessicator.

The change in reconstruction wavelength may be due to atmospheric attack. The effect of atmospheric moisture is difficult to understand, because if moisture is absorbed, one would expect the emulsion to swell thus shifting the reconstruction wavelength toward the infrared. However, detailed measurements (24) on dichromated gelatin have shown that the reconstruction wavelength for maximum diffraction efficiency shifts steadily toward shorter wavelengths as drying proceeds indicating that some emulsion shrinkage takes place. Furthermore, that this technique works at all for white light reflection holograms demands a closer consideration of the behavior of the gelatin during processing. The gelatin swells to a considerable extent during the wash, followed by rapid dehydration and shrinkage during the immersion in alcohol. Clearly this shrinkage is not enough to take up the slack introduced by the fixing bath. This suggests that the crosslinked gelatin formed during development and bleaching is dimensionally stable against removal of silver and silver compounds.

These results are the outcome of a large number of trials. It should be noted that many of the holograms gave extremely poor results although the procedures were followed precisely. In one instance a total of 20 consecutive failures occurred after a run of 20 successful holograms. These two batches were from different shipments. This would suggest that there is some variability in the gelatin.

Studies are continuing and it is encouraging to note that student undergraduate short-term projects are capable of producing interesting and useful results. We believe that if even a small expansion of our facilities were possible then graduate students and staff could pursue the more promising lines of research and development that arise from project work. At present, there is a lack of continuity because of the institute's limited budget and the difficulty in offering full-scale research facilities.

Figure 13.9. Holograms exhibition.

Exhibitions

In 1985, in an effort to bring some of the work of the Dublin Institute of Technology to the attention of the public, a holography exhibition was mounted in the college. A total of 41 holograms were presented using display arrangements as shown in Figure 13.9. The holograms were mainly of the display type including 8 on loan from private collections. In addition holographic applications resulting from student projects were highlighted. The entire exhibition was constructed by staff and students and was given very good coverage in the national press and on radio. A short television program was later broadcast showing the work of the holography laboratory.

At the invitation of Aer Lingus the Irish national airline, the exhibition was transferred to a much larger venue as a feature of the annual Young Scientists Exhibition in January 1986. It proved to be very successful drawing very large crowds. The display stands proved ideally suited for presenting holography in an area where extraneous lighting might otherwise have created difficulty (Fig. 13.10).

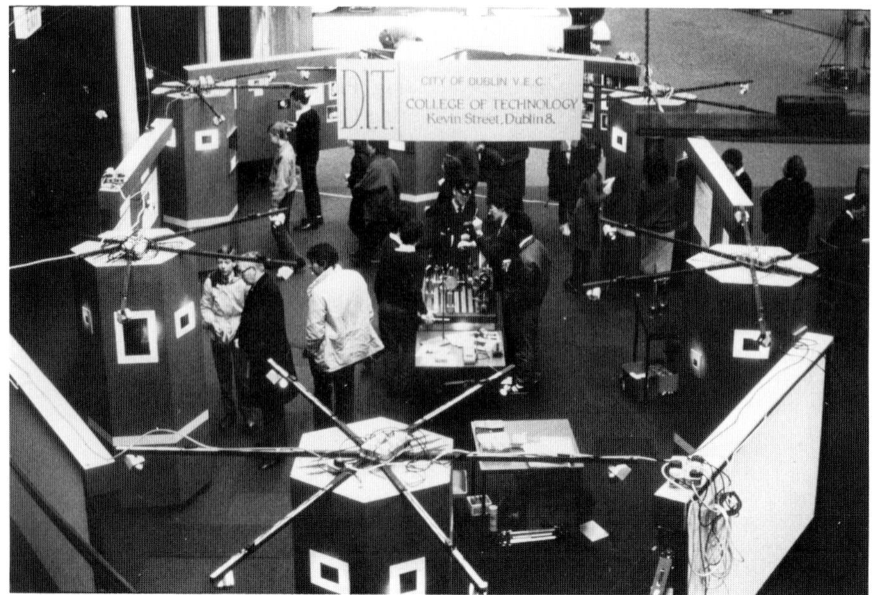

Figure 13.10. Display stands at the holographic exhibition.

Summary and conclusions

Holography has been integrated into an advanced undergraduate course in applied optics. Students are taught something of the scope and applications of Fourier optics and holographic techniques. They develop their own laboratory skills through structured experiments and project work. In this way they are better prepared for graduate study and research. Equipped with the skills developed through all of their courses, graduates are able to compete in the employment market throughout the European community.

The teaching of modern optics and holography has the built-in advantage that the subject is exciting and one can capitalize on this fact. It is worth noting that during the holography exhibitions the students appeared to have no difficulty or reticence in explaining the theory and practice of holography and communicating its exciting possibilities to the public. Although it is difficult to measure the success of any enterprise in technological education perhaps this fact may be allowed to speak for itself.

References

1. Jaffe, B. M. *Am. J. Phys.* **48,** 2, 157–8, (1980).
2. Nussbaum, A., and Phillips, R. A. *Contemporary Optics for Scientists and Engineers.* pp. 263–265, Prentice-Hall (1976).

3. Francon, M. *Modern Applications of Physical Optics.* pp. 75–78, Wiley Interscience (1963).
4. Hecht, E., and Zajac, A. *Optics.* pp. 474–481, Addison-Wesley (1974).
5. Lee, S. H. Optical Information Processing Fundamentals. *Topics in Applied Physics,* Vol. 48, pp. 43–67, Springer-Verlag (1981).
6. Ibid. 50–52.
7. Collier, R. J., Burckhardt, C. B., and Lin, L. H. *Optical Holography.* pp. 365–367, Academic Press (1971).
8. Vest, C. M. *Holographic Interferometry.* John Wiley (1979).
9. Hariharan, P. *Optical Holography.* pp. 120–128. Cambridge University Press (1984).
10. Ref. 7, pp. 368–375.
11. Smith, H. *Principles of Holography,* pp. 202–205, Wiley Interscience (1969).
12. Spierings, W. *Holosphere* **10,** 7, 1–2, (1981).
13. Smith, S. L., and Cevtkovich, *Proc. SPIE* **462,** 8–13, (1984).
14. Denisyuk, Yu. N. *Soviet Physics-Doklady* **7,** 543–545, (1962).
15. Johnson, K. M., Hesselink, L., and Goodman, J. W., *Proc. SPIE* **402,** 80–87, (1983).
16. Johnson, K. M., Hesselink, L., and Goodman, J. W. *Appl. Opt.* **23,** 2, 218–227, (1984).
17. Graver, W., Gladden, J. W., and Eastes, J. W. *Appl. Opt.* **19,** 9, 1529–1536, (1980).
18. Hariharan, P. *Appl. Opt.* **25,** 13, 2040–2042 (1986).
19. Curran, R. K., and Shankoff, T. A. *Appl. Opt.* **9,** 1651–1657, (1970).
20. McGrew, S. P. *Proc. SPIE* **215,** 24–31, (1980).
21. Samoilovich, D. M., Zeichner, A., and Freisem, A. A. *Photo. Sci. Eng.* **24,** 161–166, (1980).
22. Gaffney, J. *Silver Halide Sensitized Gelatin Holograms.* Dublin Institute of Technology (1987).
23. Solymar, L., and Cooke, D. J. *Volume Holography and Volume Gratings.* Academic Press (1981).
24. McCauley, D. G., Simpson, C. E., and Nubach, W. J., *Appl. Opt.* **12,** 232–242, (1973).

Conclusion

At the conclusion of this First International Symposium on the Industrial Uses of Holography we would like to underline the quality of the papers presented and of the discussions that followed.

Each one of us has drawn our own conclusions based on our particular interests, but it appears that we are all in agreement on a certain number of points:

1. The need for a better and more frequent communication between holographists, applied researchers, and testing engineers.
2. The creation of special groups on disciplines such as stress analysis, vibration analysis, ultrasound imaging, and optical data processing.
3. Greatly increase the teaching of holography in applied optics and material engineering courses in universities and vocational schools.
4. To promote and encourage the use of holography in remote sensing areas where the accuracy and spatial resolution are important.

The success of this Symposium illustrates the pertinence of point 1. We intend to organize this kind of event on a yearly basis.

The diversity of the highly specialized papers included in these Proceedings as well as their impact on actual problems encountered in industry, justify the concerns stated in point 2.

The teaching of holography is still lacking in many educational institutions where this new technique should play a major role as a necessary complement to other disciplines. This is particularly true in Material Sciences. A comprehensive program is offered by the Applied Optics Laboratories of New Mexico State University in association with local industries as well as government agencies, thus creating the proper environment for applications oriented students.

After a slow start due to the lack of reliable laser sources and recording materials, holography is now rapidly becoming widely accepted and in most cases preferred optical diagnosis tool. We should, therefore, congratulate the attendants to this 1987 Symposium for their major contribution to the promotion of Industrial Holography.

Index